NEW CALCULATIONS IN CHEMISTRY

Student problems at 16+

D. G. DAVIES, B Sc. M.R.I.C.
Vice Principal, Palmer's College

T. V. G. KELLY, B Sc.
Head of Curriculum Support Services, William Ellis School

Bell & Hyman

PREFACE

One of the surest methods of helping students to understand chemical concepts of a quantitative nature is to have them perform a series of well chosen calculations or data analyses. Course textbooks rarely contain a sufficient number of questions of this type for the teacher to make his or her own selection. It is hoped that this collection of questions will go some way towards remedying this.

The majority of the questions have been composed by the authors with the new syllabuses in mind, and 'new' topics at this level (e.g. reaction rates, thermochemistry, the mole concept, cell e.m.fs.) have been given much space. Many questions require graphical methods to be used for the expression and analysis of experimental data. In general, each section has been arranged so that the first few questions familiarise students with the basic ideas and calculation routines.

Chemical nomenclature and physical units employed are largely those recommended by IUPAC—a list of abbreviations is given on p. iv.

We would like to thank Dr. D. J. Waddington for his continued advice and encouragement and our publishers for their kindness and consideration at all stages in the preparation of this book.

D.G.D.
T.K.

ACKNOWLEDGEMENTS

We would like to thank the following examining boards for their kind permission to reprint questions from past examination papers at Ordinary Level:

University of London Entrance and School Examinations Council (L)
Oxford Delegacy of Local Examinations (O)
University of Cambridge Local Examinations Syndicate (C)
Oxford and Cambridge Schools Examination Board (O & C)

In some cases abbreviations, units and chemical nomenclature have been altered to bring them in line with current practice. This in no way alters the substance of the questions. Alteration of units, etc., is entirely the responsibility of the authors. Numerical answers to questions which are the copyright of the above boards are the results of work by the authors, and the boards accept no responsibility for their accuracy.

First published as 'New Calculations in Chemistry for Ordinary Level' 1971 by
Mills & Boon Limited, London

Reprinted 1971
Reprinted with revisions 1975
Reprinted 1976
Reprinted 1978
Reprinted as 'New Calculations in Chemistry' 1985

Printed and bound in Great Britain by The Bath Press, Avon

CONTENTS

                                                                            Page

        ABBREVIATIONS   iv

        ATOMIC NUMBERS AND APPROXIMATE ATOMIC WEIGHTS   iv

Section 1   THE MOLE (g-atom, g-molecule, g-formula weight, etc.)   1

     2   THE GAS LAWS   4

     3   MOLAR VOLUMES AND RELATIVE (VAPOUR) DENSITY   5

     4   REACTING WEIGHTS (not including gas volumes)   9

     5   REACTING WEIGHTS (including gas volumes)   12

     6   REACTING VOLUMES OF GASES   15

     7   STANDARD SOLUTIONS   19

     8   THERMOCHEMISTRY   25

     9   ELECTROCHEMISTRY   34

   10   DEDUCTION OF EQUATIONS   39

   11   DIFFUSION AND EFFUSION OF GASES   43

   12   RATES OF REACTION   45

   13   SOLUBILITY AND SOLUBILITY CURVES   52

        ANSWERS   56

ABBREVIATIONS

A	ampere
at. wt.	atomic weight
aq	water
(aq)	aqueous solution
atm	atmosphere
C	coulomb
°C	degree Celsius (centigrade)
cal (kcal)	calorie (kilocalorie)
cm	centimetre
cmHg	centimetre of mercury pressure
dm	decimetre
e	electron
E	electromotive force
E°	standard electrode potential
F	Faraday constant
g (kg)	gramme (kilogramme)
(g)	gaseous
g—eq	gramme—equation
h	hour
ΔH	enthalpy (heat) change
J (kJ)	joule (kilojoule)
K	Kelvin
L	Avogadro number (constant)
(l)	liquid
M	molar
min	minute
mm	millimetre
mmHg	millimetre of mercury pressure
mol	mole
mol. wt.	molecular weight
s	second
(s)	solid
s.t.p.	standard temperature and pressure
V	volt
Z	atomic (proton) number

ATOMIC NUMBERS AND APPROXIMATE ATOMIC WEIGHTS

Element	Symbol	Atomic Number	Approximate Atomic Weight
Aluminium	Al	13	27
Antimony	Sb	51	122
Argon	Ar	18	40
Barium	Ba	56	137.5
Bromine	Br	35	80
Caesium	Cs	55	133
Calcium	Ca	20	40
Carbon	C	6	12
Cerium	Ce	58	140
Chlorine	Cl	17	35.5
Chromium	Cr	24	52
Copper	Cu	29	63.5
Fluorine	F	9	19
Gold	Au	79	197
Helium	He	2	4
Hydrogen	H	1	1
Iodine	I	53	127
Iron	Fe	26	56
Krypton	Kr	36	84
Lead	Pb	82	207
Lithium	Li	3	7
Magnesium	Mg	12	24
Manganese	Mn	25	55
Mercury	Hg	80	201
Neon	Ne	10	20
Nickel	Ni	28	59
Nitrogen	N	7	14
Oxygen	O	8	16
Phosphorus	P	15	31
Platinum	Pt	78	195
Potassium	K	19	39
Radon	Rn	86	222
Rubidium	Rb	37	85.5
Silicon	Si	14	28
Silver	Ag	47	108
Sodium	Na	11	23
Sulphur	S	16	32
Tin	Sn	50	119
Titanium	Ti	22	48
Xenon	Xe	54	131
Zinc	Zn	30	65.5

1
THE MOLE
(g-atom, g-molecule, g-formula weight, etc.)

1. Calculate the molecular weights (or ionic, or formula weights) of the following species:

(a) O_2 (b) CO_2 (c) $SO_4{}^{2-}$

(d) KNO_3 (e) C_2H_5OH (f) $FeSO_4, 7H_2O$

2. Calculate the number of moles (gramme-molecules, gramme-atoms, gramme-formulae, gramme-ions) in the weights of the given substances

(a) $128 \text{ g } O_2$ (b) $22 \text{ g } CO_2$ (c) $58.5 \text{ g } NaCl$

(d) $25.25 \text{ g } KNO_3$ (e) $120 \text{ g } Na_2SO_4$ (f) $414 \text{ g } C_2H_5OH$

3. How many moles of

(a) Al^{3+} are there in 2 moles $Al_2(SO_4)_3$

(b) $SO_4{}^{2-}$ are there in 0.7 mole $CuSO_4$

(c) $PO_4{}^{3-}$ are there in 0.2 mole $Ca_3(PO_4)_2$

(d) NH_3 are there in 0.05 mole $Ag(NH_3)_2Cl$?

4. How many moles of ions (i.e. total of all species) are there in

(a) 0.2 mole $FeSO_4, 7H_2O$

(b) 1.5 moles $Cr_2(SO_4)_3$

(c) 20 g NaOH

(d) 10 g $CaCO_3$?

5. How many grammes of material are there in each of the following?

(a) 3 moles H_2 (b) 0.1 mole HCl

(c) 0.04 mole CO_2 (d) 10^{-4} mole $CH_3(CH_2)_{10}CH_3$

(e) 0.9 mole K_2SO_4 (f) 2 moles H_2S

(g) 0.2 mole $CaCO_3$ (h) 1 mole OH^-

6. Calculate the weight, in grammes, of the first element having the same number of atoms as the given weight of the second element.

	First element	Second element
(a)	Ag	63.5 g Cu
(b)	Mg	23.0 g Na
(c)	K	20.7 g Pb
(d)	Fe	2.6 g Cr
(e)	Al	131.0 g Zn

7. Calculate the number of moles of the first substance having the same number of molecules as the stated weight of the second substance.

	First substance	Second substance
(a)	CO_2	32.0 g SO_2
(b)	HBr	73.0 g HCl
(c)	Cl_2	25.4 g I_2
(d)	Ar	1.6 g O_2
(e)	C_2H_4	3.9 g C_6H_6

8. 42 g of iron (Fe) react with 16 g of oxygen (O_2) to produce an oxide of iron (Fe_3O_4)

    (a) Rewrite the above sentence in terms of moles.
    (b) How many moles of Fe_3O_4 are formed?
    (c) What is the simplest ratio of
$$\text{moles Fe : moles } O_2 \text{ : moles } Fe_3O_4$$
involved in the reaction?

9. (a) $2Al(s) + 6HCl(aq) \rightarrow 2AlCl_3(aq) + 3H_2(g)$

How many moles of (i) $AlCl_3$, (ii) H_2 are produced when 0.5 mole of Al reacts with excess HCl?

    (b) $2Pb(NO_3)_2(s) \rightarrow 2PbO(s) + 4NO_2(g) + O_2(g)$

How many moles of (i) O_2, (ii) PbO, (iii) NO_2 are produced when 6.2 moles of lead (II) nitrate are heated until decomposition is complete?

10. 11.95 g of an oxide of lead, on reduction, gave 10.35 g of the metal. How many atoms of oxygen are combined with one atom of lead in this oxide?   (O & C)

11. Given below are the weight compositions of a number of compounds. Work out the simplest (empirical) formula for each of them.

(a)	P	22.5%	Cl	77.5%		
(b)	Ca	40.0%	C	12.0%	O	48.0%
(c)	Pb	86.5%	O	13.5%		
(d)	C	92.4%	H	7.6%		
(e)	N	26.2%	H	7.5%	Cl	66.3%
(f)	Na	29.1%	S	40.5%	O	30.4%
(g)	C	82.8%	H	17.2%		

12. In an experiment to determine the formula of hydrated sodium sulphate, Na_2SO_4, xH_2O, the following results were obtained:

Weight of dish	= 15.14 g
Weight of dish + hydrated salt	= 18.36 g
Weight of dish + anhydrous salt after heating to constant weight	= 16.56 g

What is the formula of hydrated sodium sulphate?

13. A hydrate of zinc sulphate, $ZnSO_4$, xH_2O, contained 22.8% of zinc by weight. Find its formula.

14. The atomic weight of a trivalent metal is 27. Calculate the percentage of metal in its oxide.   (O & C)

15. The atomic weight of a bivalent metal is 9. Calculate the percentage of metal in its chloride.

16. Calculate the percentage by weight of
(a) lead in lead (IV) oxide, PbO_2
(b) nitrogen in ammonium sulphate, $(NH_4)_2SO_4$
(c) copper in copper (II) sulphate crystals, $CuSO_4$, $5H_2O$
(d) chlorine in common salt, NaCl
(e) iron in haematite ore, Fe_2O_3

17. Calculate the percentage by weight of water of crystallisation in the following salt hydrates:
(a) washing soda, Na_2CO_3, $10H_2O$
(b) copper (II) sulphate crystals, $CuSO_4$, $5H_2O$
(c) iron (II) sulphate crystals, $FeSO_4$, $7H_2O$
(d) Epsom salt $MgSO_4$, $7H_2O$
(e) Alum, K_2SO_4, $Al_2(SO_4)_3$, $24H_2O$

2
THE GAS LAWS

1. What are the following temperatures on the Kelvin scale? (0°C = 273 K)

(a)　17°C   (b)　25°C   (c)　-13°C

(d) -200°C   (e)　300°C   (f)　-29°C

2. What are the following temperatures on the Celsius (centigrade) scale? (0°C = 273 K)

(a) 300 K   (b) 212 K   (c) 271 K

(d)　50 K   (e) 654 K   (f)　0 K

3. Convert the following gas volumes from s.t.p. (0°C and 760 mm of mercury pressure) to the stated temperature and pressure.

	Volume at s.t.p.	Convert to	
(a)	55.0 cm^3	15°C	750 mmHg
(b)	22.40 dm^3	25°C	760 mm Hg
(c)	28.3 cm^3	-10°C	765 mmHg
(d)	5.40 dm^3	100°C	2 atm

4. Convert the following gas volumes, at the temperatures and pressures indicated, to s.t.p.

	Volume	Temperature	Pressure
(a)	35 cm^3	15°C	760 mmHg
(b)	113 cm^3	-20°C	740 mmHg
(c)	50 dm^3	500°C	200 atm
(d)	25.20 cm^3	-6°C	765 mmHg

5. The following table gives gas volumes at the temperatures and pressures at which they were measured. Calculate the volumes that the gases would occupy if the conditions were changed to those indicated.

	Volume	Measured at		Convert to	
(a)	295 cm^3	22°C	765 mmHg	44°C	765 mmHg
(b)	96.5 cm^3	273°C	42 cmHg	0°C	84 mmHg
(c)	380 cm^3	27°C	760 mmHg	77°C	760 mmHg
(d)	2.6 dm^3	100°C	320 mmHg	15°C	755 mmHg

3
MOLAR VOLUMES AND RELATIVE (VAPOUR) DENSITY

Molar volumes

1. For each of the following gases calculate the volume, in dm³, occupied by 1 mole of the gas at s.t.p. You are given the molecular weight of the gas and its density at s.t.p. in grammes per litre.

Gas	Mol. wt.	Density at s.t.p. (g dm⁻³)
(a) hydrogen	2	0.0893
(b) oxygen	32	1.429
(c) nitrogen	28	1.250
(d) carbon dioxide	44	1.964
(e) ammonia	17	0.759

(f) What general conclusion do you reach?

2. What is the volume at s.t.p. of
(a) 71 g of chlorine, Cl_2
(b) 15 g of nitrogen monoxide, NO
(c) 68 g of hydrogen sulphide, H_2S
(d) 6.4 g of sulphur dioxide, SO_2
(e) 0.7 g of carbon monoxide, CO?

3. Using the fact that 1 mole of any gas at s.t.p. occupies 22.4 dm^3, calculate the molecular weights of the following gases.

Weight of gas		Volume of gas
(a)	0.15 g	112 cm^3 at s.t.p.
(b)	0.85 g	0.56 dm^3 at s.t.p.
(c)	0.18 g	250 cm^3 at s.t.p.
(d)	2.86 g	1 dm^3 at s.t.p.
(e)	0.46 g	244 cm^3 at 76 cmHg and 25°C

4. (a) The weight of 1 dm^3 of gas at s.t.p. is 1.52 g. What is its molecular weight?

(b) By how many grammes is 5.6 litres of sulphur dioxide heavier than the same volume of carbon dioxide (both gases at s.t.p.)? (C)

5. When 4.73 g of a solid was heated the residue weighed 4.10 g and 320 cm^3 of a gas (measured at s.t.p.) was evolved. Calculate the molecular weight of the gas. (L)

6. The action of an acid displaced from a solid 733 cm^3 of a gas measured at 17°C and 740 mmHg. The apparatus was found to have lost 1.025 g in weight after the evolution of the gas. Calculate the molecular weight of the gas. (C)

7. Calculate the gramme-atomic (molar) volumes of the following solid elements. Their densities (g cm^{-3}) and atomic weights are given.

Element	at. wt.	Density (g cm^{-3})
(a) Aluminium	27	2.70
(b) Barium	137	3.51
(c) Lead	207	11.3
(d) Magnesium	24	1.74
(e) Nickel	59	8.90

8. Calculate the atomic volumes of the following alkali metals, you are given their atomic numbers (Z), atomic weights (A) and their densities.

Element	Z	A	Density (g cm^{-3})
(a) Lithium, Li	3	6.9	0.54
(b) Sodium, Na	11	23.0	0.97
(c) Potassium, K	19	39.1	0.87

The atomic volumes of the other alkali metals are Rubidium, Rb (Z = 37) 55.7 cm^3 Caesium, Cs (Z = 55) 71.0 cm^3.

Using these values and those calculated above, plot a graph of atomic volume against atomic number. Explain the shape of the graph.

9. Calculate the molar (g-molecular) volumes of the following compounds.

Compound	Formula	Density (g cm^{-3})
(a) Tetrachloromethane	CCl_4	1.59
(b) Ethanol	C_2H_5OH	0.79
(c) Water	H_2O	1.00
(d) Iodomethane	CH_3I	2.28

10. The densities of the chlorides of the alkali metals are as follows:

Chloride	LiCl	NaCl	KCl	RbCl	CsCl
Density (g cm^{-3})	2.07	2.16	1.98	2.80	3.99

(a) Calculate the g-formula (molar) volumes of these chlorides.

(b) Look up the atomic numbers of the metals.
Plot a graph of 'g-formula volume of chloride' against 'atomic number of metal'. Comment briefly on your graph.

11. The halides of sodium have the following formulae and densities:

Halide	NaF	NaCl	NaBr	NaI
Density (g cm^{-3})	2.56	2.16	3.20	3.67

(a) Calculate the g-formula (molar) volumes of these halides.

(b) Look up the atomic numbers of the halogens.
Plot a graph of 'g-formula volume of sodium halide' against 'atomic number of halogen'.
Explain the shape of the graph.

12. The following list of compounds (their densities are also given) is comprised of hydrocarbons called alkanes.
Calculate and tabulate their molar volumes.
Plot a graph of 'molar volume' against 'number of carbon atoms' for this series of alkanes.

7

Compound	Formula	Density (g cm^{-3})
(a) pentane	$CH_3(CH_2)_3CH_3$	0.626
(b) hexane	$CH_3(CH_2)_4CH_3$	0.660
(c) heptane	$CH_3(CH_2)_5CH_3$	0.684
(d) octane	$CH_3(CH_2)_6CH_3$	0.703
(e) nonane	$CH_3(CH_2)_7CH_3$	0.718

What do you conclude from your graph?

Relative (vapour) density

13. Calculate the relative (vapour) densities of the following gases:

(a) sulphur dioxide, SO_2 (b) carbon monoxide, CO
(c) carbon dioxide, CO_2 (d) ammonia, NH_3
(e) hydrogen sulphide, H_2S (f) chlorine, Cl_2
(g) argon, Ar   (h) hydrogen iodide, HI
(i) air — which may be considered
to be a mixture of 20% oxygen (O_2)
and 80% nitrogen (N_2) by volume.

14. In the following table are given the empirical (simplest) formulae of a number of compounds, together with their relative (vapour) densities. What are their molecular formulae?

Empirical (simplest) formula		Relative density
(a)	CH_2Br	94
(b)	CH	39
(c)	CH_2O	15
(d)	CH_2O	30
(e)	SCl	67.5

15. A gaseous hydrocarbon has a formula of the type C_nH_{2n+2}. The relative (vapour) density of this gas was found experimentally and the results were as follows:

Weight of evacuated flask = 26.31 g
Weight of flask full of hydrogen = 26.52 g
Weight of flask full of gas at the same temperature and pressure = 32.40 g

Calculate (a) the relative (vapour) density of the hydrocarbon, (b) its molecular weight. (c) What is its molecular formula?

4
REACTING WEIGHTS
(not including gas volumes)

1. Calculate the weight of the residue left on heating:

(a) 1.56 g of silver oxide to leave silver
$$2Ag_2O(s) \rightarrow 4Ag(s) + O_2(g)$$

(b) 6.25 g of copper (II) carbonate to leave copper (II) oxide
$$CuCO_3(s) \rightarrow CuO(s) + CO_2(g)$$

(c) 2.67 g of lead (II) carbonate to leave lead (II) oxide
$$PbCO_3(s) \rightarrow PbO(s) + CO_2(g)$$

(d) 5.00 g of copper (II) sulphate crystals to leave copper (II) oxide
$$CuSO_4, 5H_2O(s) \rightarrow CuO(s) + SO_3(g) + 5H_2O(g)$$

(e) 5.56 g of iron (II) sulphate crystals to leave iron (III) oxide
$$2[FeSO_4, 7H_2O(s)] \rightarrow Fe_2O_3(s) + SO_2(g) + SO_3(g) + 14H_2O(g)$$

2. What weight of ammonia is evolved when 2.00 g of ammonium sulphate is boiled with excess sodium hydroxide solution?
$$(NH_4)_2SO_4(aq) + 2NaOH(aq) \rightarrow Na_2SO_4(aq) + 2NH_3(g) + 2H_2O(l)$$

3. What weight of lead would be obtained on complete reduction of 1.60 g of lead (IV) oxide by hydrogen?
$$PbO_2(s) + 2H_2(g) \rightarrow Pb(s) + 2H_2O(g)$$

4. 50 cm^3 of an iron (III) chloride solution were treated with an excess of sodium hydroxide solution when iron (III) hydroxide was precipitated
$$FeCl_3(aq) + 3NaOH(aq) \rightarrow Fe(OH)_3(s) + 3NaCl(aq)$$

This precipitate, after washing, was heated to constant weight to yield 0.80 g of iron (III) oxide
$$2Fe(OH)_3(s) \rightarrow Fe_2O_3(s) + 3H_2O(g)$$

Calculate the concentration of iron (III) chloride in g dm^{-3} in the original solution.

5. A certain iron ore is found to contain 85% of iron (III) oxide (Fe_2O_3) and no other iron-containing substance. Calculate how many tons of ore are required to produce 1 ton of iron, assuming the process to be 100% efficient. (L)

6. What weight of sulphuric acid could be produced from 1 ton of sulphur by the Contact Process, assuming the total conversion to be 90% efficient?

7. A piece of marble (calcium carbonate, $CaCO_3$) weighing 14.3 g was placed in 100 cm^3 of a solution containing 63.0 g of nitric acid (HNO_3) per dm^3. What weight of marble remained when reaction was complete?

8. Ammonium dichromate decomposes on heating:
$$(NH_4)_2Cr_2O_7(s) \rightarrow Cr_2O_3(s) + N_2(g) + 4H_2O(g)$$
What would be the weight of the residue remaining after heating 12.6 g of this compound? How many moles of nitrogen were evolved?

9. Ammonium nitrate is manufactured by neutralising nitric acid with ammonia gas. The equation for the reaction is
$$NH_3(g) + HNO_3(aq) \rightarrow NH_4NO_3(aq)$$
What weight of (i) ammonia, (ii) nitric acid would be required to make 400 tons of ammonium nitrate?

10. What is the maximum weight of hydrated calcium chloride of formula $CaCl_2$, $6H_2O$ which can be made by dissolving 5 g of calcium carbonate ($CaCO_3$) in dilute hydrochloric acid? (O & C)

11. The reaction between silver nitrate solution and zinc chloride solution is
$$2AgNO_3(aq) + ZnCl_2(aq) \rightarrow 2AgCl(s) + Zn(NO_3)_2(aq)$$
a precipitate of silver chloride being formed.

Calculate (i) the maximum weight of silver chloride which could be precipitated from a solution containing 8.5 g of silver nitrate, and (ii) the weight of zinc chloride involved in this precipitation.

12. 2.60 g of a mixture of iron and iron (III) oxide (Fe_2O_3) on complete reduction left 2.12 g of iron. Calculate the percentage composition of the mixture.

13. 6.00 g of a mixture of zinc oxide and zinc carbonate were heated to constant weight. The residue of zinc oxide weighed 4.85 g. Calculate the percentage composition of the mixture.

14. 5.50 g of a mixture of the anhydrous salts sodium nitrate ($NaNO_3$) and sodium sulphate (Na_2SO_4) were dissolved in water and then treated with excess barium chloride solution. The precipitate of barium sulphate ($BaSO_4$) which was formed weighed 8.22 g. Calculate the percentage composition of the mixture.

15. 36 g of a mixture of copper with magnesium were found to displace 2 g of hydrogen from dilute sulphuric acid. What was the percentage by weight of copper in the mixture?

(O)

16. 5.04 g of a mixture of anhydrous sodium carbonate and sodium hydrogen carbonate when heated at a moderate temperature gave a constant weight residue of 4.11 g. Calculate the percentage of anhydrous sodium carbonate in the mixture. What would be the loss in weight when 1.00 g of the same mixture is treated with excess of a non-volatile acid?

17. The maximum weight of copper (II) sulphate crystals, $CuSO_4$, $5H_2O$ obtainable from 20.0 g of a sample of brass containing only zinc and copper is 37.2 g. What weight of zinc was present in the 20.0 g of brass?

(C)

18. 3.10 g of sodium carbonate Na_2CO_3, xH_2O were dissolved in water. To the resulting solution excess calcium chloride solution was added when calcium carbonate was precipitated

$$Na_2CO_3 (aq) + CaCl_2 (aq) \rightarrow CaCO_3 (s) + 2NaCl(aq)$$

This precipitate was found to weigh 2.50 g. What was the formula of the hydrate?

19. A mixture of anhydrous sodium carbonate and sodium hydrogen carbonate, on treatment with excess acid, yielded 0.03 mole of carbon dioxide. A similar weight of this mixture on heating gave 0.01 mole of the same gas. Calculate the mole ratio $Na_2CO_3 : NaHCO_3$ in the mixture. Both react with acid yielding carbon dioxide. Sodium carbonate does not decompose on heating but the sodium hydrogen carbonate does:

$$2NaHCO_3 (s) \rightarrow Na_2CO_3 (s) + CO_2 (g) + H_2O(g)$$

20. When 2.50 g of a copper ore were heated, 1.73 g of copper (II) oxide were left. Which of the following formulae best represents the composition of this ore?

$2Cu(OH)_2$, $CuCO_3$ $CuCO_3$ $Cu(OH)_2$, $2CuCO_3$

21. (a) A crystalline substance was analysed with the following results. 1.000 g of it was heated gently to constant weight; the residue weighed 0.703 g. The gas evolved was passed through a silica-gel drying tube which gained in weight by 0.199 g and then through potassium hydroxide which gained 0.098 g. Which of the following formulae best represents the composition of the crystals?

Na_2CO_3, H_2O or Na_2CO_3, $NaHCO_3$, $2H_2O$ or $NaHCO_3$

(b) The 0.703 g residue was then treated with excess of dilute hydrochloric acid in an apparatus which allowed carbon dioxide, but no other substance to escape. Calculate the loss in weight which would be suffered by this apparatus.

(L)

5
REACTING WEIGHTS
(including gas volumes)

1. Calculate the volume of gas, expressed at s.t.p., evolved when the following solids are heated:

(a) 2.32 g of silver oxide, which evolves oxygen

$$2Ag_2O(s) \rightarrow 4Ag(s) + O_2(g)$$

(b) 13.35 g of lead (II) carbonate which evolves carbon dioxide

$$PbCO_3(s) \rightarrow PbO(s) + CO_2(g)$$

(c) 11.95 g of lead (IV) oxide which evolves oxygen

$$2PbO_2(s) \rightarrow 2PbO(s) + O_2(g)$$

(d) 10.00 g of zinc carbonate which evolves carbon dioxide

$$ZnCO_3(s) \rightarrow ZnO(s) + CO_2(g)$$

2. Calculate the volume of gas produced (and measured under the stated conditions) when each of the solids below is treated as indicated. All steam evolved may be assumed to condense completely.

(a) 12.60 g ammonium dichromate, heated; gas collected at $15°C$ and 760 mmHg pressure

$$(NH_4)_2Cr_2O_7(s) \rightarrow Cr_2O_3(s) + N_2(g) + 4H_2O(l)$$

(b) 2.17 g mercury (II) oxide, heated; gas collected at $25°C$ and 760 mmHg pressure.

$$2HgO(s) \rightarrow 2Hg(l) + O_2(g)$$

(c) 15.0 g marble ($CaCO_3$), treated with excess hydrochloric acid; gas collected at $18°C$ and 765 mmHg pressure.

$$CaCO_3(s) + 2HCl(aq) \rightarrow CaCl_2(aq) + CO_2(g) + H_2O(l)$$

(d) 16.0 g ammonium nitrate, boiled with excess sodium hydroxide solution; gas collected at $10°C$ and 740 mmHg pressure.

$$NH_4NO_3(aq) + NaOH(aq) \rightarrow NaNO_3(aq) + NH_3(g) + H_2O(l)$$

(e) 8.0 g of sulphur, burned in excess oxygen; gas collected at $50°C$ and 780 mmHg pressure.

$$S(s) + O_2(g) \rightarrow SO_2(g)$$

3. 10.6 g of anhydrous sodium carbonate were treated with excess hydrochloric acid
$$Na_2CO_3 (s) + 2HCl(aq) \rightarrow 2NaCl(aq) + CO_2 (g) + H_2O(l)$$

(a) What volume of carbon dioxide, expressed at s.t.p., was produced? (b) What weight of solid would have been left had the resulting solution been evaporated to dryness?

4. Potassium permanganate decomposes on heating
$$2KMnO_4 (s) \rightarrow K_2MnO_4 (s) + MnO_2 (s) + O_2 (g)$$

What weight of this salt would be required to produce 1 dm^3 of oxygen, measured at s.t.p? What would be the weight of the residue left after the heating?

5. Calculate the volume of carbon monoxide (measured at s.t.p.) needed to reduce 111.5 g of litharge (PbO) to lead. (L)

6. Carbon dioxide and strongly heated carbon react to produce carbon monoxide according to the equation
$$CO_2 (g) + C(s) \rightarrow 2CO(g)$$

What volume of carbon monoxide (measured in dm^3 at s.t.p.) can be produced in this way from 3 g of carbon? (O & C)

7. From the equation
$$4FeS_2 (s) + 11O_2 (g) \rightarrow 2Fe_2O_3 (s) + 8SO_2 (g)$$

calculate (i) the volume of sulphur dioxide, measured at s.t.p., that would be obtained by the complete oxidation of 1 kg of iron pyrites and (ii) the volume of air, measured at s.t.p., used in the oxidation, assuming that air contains 20% oxygen by volume. (C)

8. Potassium chlorate when heated gives off all its oxygen, whereas potassium nitrate when heated gives off only part of its oxygen.
    Write equations for these decompositions.
    Calculate the relative weights of these salts which give off equal volumes of oxygen, the gas being collected at the same temperature and pressure in each case. (L)

9. 4.00 g of ammonium nitrate were gently heated till no residue was left. The gases evolved from this decomposition
$$NH_4NO_3 (s) \rightarrow N_2O(g) + 2H_2O(g)$$

were passed through a series of U-tubes containing anhydrous calcium chloride and the dry dinitrogen monoxide was then collected. What was (a) the increase in weight of the U-tubes, (b) the volume of dinitrogen monoxide collected, expressed at s.t.p?

10. 0.715 g of hydrated sodium carbonate (Na_2CO_3, xH_2O) was treated with excess nitric acid, and the evolved carbon dioxide collected in a syringe and measured. This volume, corrected to s.t.p., was 56 cm^3. Calculate the value of x. The equation for the reaction is
$$Na_2CO_3 (s) + 2HNO_3 (aq) \rightarrow 2NaNO_3 (aq) + CO_2 (g) + H_2O(l)$$

11. 1.32 g of ammonium sulphate was heated with excess calcium hydroxide when ammonia was evolved

$$(NH_4)_2SO_4(s) + Ca(OH)_2(s) \rightarrow CaSO_4(s) + 2NH_3(g) + 2H_2O(l)$$

The ammonia was then passed over excess heated copper (II) oxide

$$3CuO(s) + 2NH_3(g) \rightarrow 3Cu(s) + N_2(g) + 3H_2O(l)$$

Calculate the volume of nitrogen produced (a) expressed at s.t.p., (b) expressed at $15°C$ and 755 mmHg pressure. (c) How many moles of copper were formed from the copper (II) oxide?

12. Polystyrene decomposes on heating to yield the monomer, styrene, which has the formula $C_6H_5CH:CH_2$. What volume of gas at $273°C$ and 1 atmosphere pressure would be expected for each gramme of polystyrene decomposed?   (L)

13. Ethylene (C_2H_4) polymerises under certain conditions to form 'polythene' $(CH_2)_n$. What volume of ethylene gas, at $0°C$ and 76 cmHg pressure, would be required to produce 10 g of polythene?

14. Nitrogen combines with certain metals, forming compounds called nitrides; two such metals are lithium (Li) and calcium. These nitrides react with water, giving off ammonia and leaving a metallic hydroxide.

When 1.75 g of lithium nitride are treated with water, 1.18 dm^3 of ammonia at $17°C$ and 765 mmHg pressure are evolved.
(a) Calculate the empirical formula of lithium nitride.
(b) What is the valency of lithium in this compound?
(c) Write the equation for the action of water on lithium nitride.
(d) Write the formula of calcium nitride.   (L)

15. When magnesium is heated in nitrogen it forms magnesium nitride (Mg_3N_2); the latter compound reacts with water to form magnesium hydroxide $(Mg(OH)_2)$ and ammonia. Write equations for these two reactions and calculate (i) the volume of nitrogen, measured at $17°C$ and 750 mmHg, which would be needed to react with 1.2 g of magnesium, and (ii) the volume of ammonia, at the same temperature and pressure, which would be obtained from the magnesium nitride so obtained.   (L)

16. 'White lead' may be regarded as a compound of lead (II) carbonate $(PbCO_3)$ and lead (II) hydroxide $(Pb(OH)_2)$. When 0.775 g of it is heated strongly, 0.018 g of water and 44.8 cm^3 (at s.t.p.) of carbon dioxide are evolved and a residue of lead monoxide is left. Deduce the formula of white lead.   (L)

6
REACTING VOLUMES OF GASES

1. In each of the following reactions the volume of one of the gases involved in the reaction is given. Calculate the volumes of the other gases which either react or are produced. All gas volumes are measured at the same temperature and pressure.

(a)
$$H_2(g) + Cl_2(g) \rightarrow 2HCl(g)$$
$$1.0 \text{ dm}^3$$

(b)
$$2NO(g) + O_2(g) \rightarrow 2NO_2(g)$$
$$10 \text{ cm}^3$$

(c)
$$2H_2S(g) + 3O_2(g) \rightarrow 2SO_2(g) + 2H_2O(g)$$
$$100 \text{ cm}^3$$

(d)
$$CO_2(g) + C(s) \rightarrow 2CO(g)$$
$$50 \text{ cm}^3$$

(e)
$$2C_4H_{10}(g) + 13O_2(g) \rightarrow 8CO_2(g) + 10H_2O(g)$$
$$208 \text{ cm}^3$$

(f)
$$2H_2S(g) + SO_2(g) \rightarrow 3S(s) + 2H_2O(l)$$
$$15 \text{ cm}^3$$

2. A fuel gas contains (by volume) 50% hydrogen (H_2) and 44% carbon monoxide (CO), the other 6% being incombustible.
(a) Write equations for the combustion of hydrogen and carbon monoxide.
(b) Calculate the volume of oxygen required to burn 50 cm^3 of this fuel gas.
(c) What volume of air would be required if air contained 20% oxygen by volume?
   All gases are at the same temperature and pressure.

3. To a mixture of 20 cm^3 of carbon monoxide and 30 cm^3 of methane, 100 cm^3 of oxygen were added and a spark was passed.
   Assuming constant pressure throughout in each case, calculate the composition of the resulting gas if all the measurements are made (i) at room temperature, (ii) at 110°C.
   Methane reacts with oxygen according to the equation   (L)
$$CH_4(g) + 2O_2(g) \rightarrow CO_2(g) + 2H_2O(g)(1)$$

4. Calor gas consists of 95% butane (C_4H_{10}) and 5% pentane (C_5H_{12}) by volume. Assuming that air contains 20% of oxygen by volume, calculate the volume of air needed for the complete combustion of 100 dm^3 of Calor gas. All measurements are made at the same temperature and pressure. (L)

5. A mixture of 40 cm^3 of oxygen and 40 cm^3 of hydrogen was exploded by sparking. What was the composition and volume of the gas left? All volumes were measured at the same room temperature and pressure.

6. The gas formaldehyde, CH_2O, burns completely to carbon dioxide and water. Write down the volume of oxygen required to burn 100 cm^3 of formaldehyde. Both volumes are measured at the same temperature and pressure. (O & C)

7. 10 cm^3 of a mixture of carbon monoxide and nitrogen required 2.5 cm^3 of oxygen for complete combustion. Calculate the percentage of nitrogen in the mixture. (O & C)

8. Semi-water gas may be considered as containing 20% hydrogen, 30% carbon monoxide and 50% nitrogen.
  If 40 cm^3 of this gas are exploded with 20 cm^3 of oxygen, what will be the composition of the final mixture of gases, all volumes being measured at room temperature and pressure? (O)

9. 20 cm^3 of a gaseous mixture of nitrogen and ammonia were passed repeatedly over heated copper (II) oxide using syringes. When no further volume change occurred the volume of the residual gas was 16 cm^3. What was the composition of the mixture? Nitrogen does not react with copper (II) oxide but ammonia does:
$$3CuO(s) + 2NH_3(g) \rightarrow 3Cu(s) + N_2(g) + 3H_2O(l)$$
  Both gas volumes were measured at the same room temperature and pressure.

10. 200 cm^3 of a mixture of nitrogen and nitrogen monoxide (NO) were contained in a syringe. This mixture was passed repeatedly over heated lead until no further change occurred. The final volume was found to be 110 cm^3. Calculate the percentage composition, by volume, of the original mixture. Nitrogen monoxide reacts with lead according to the equation
$$2NO(g) + 2Pb(s) \rightarrow 2PbO(s) + N_2(g)$$
Nitrogen does not react with lead. All volumes were measured at the same temperature and pressure.

11. 24.0 cm^3 of air were mixed with 15.0 cm^3 (an excess) of hydrogen and the mixture was exploded. The final volume of gas was 24.0 cm^3. All volumes were measured at 15°C and 765 mmHg pressure. Calculate the percentage, by volume, of oxygen in the air.

16

12. Water gas may be considered to be a mixture of hydrogen and carbon monoxide. 60 cm^3 of this gas were mixed with excess oxygen and exploded. The products, after cooling to the original temperature, were treated with excess sodium hydroxide solution when the volume decreased by 30 cm^3.
(a) What was the percentage composition by volume of the water gas?
(b) What volume of oxygen would be necessary for the complete combustion of 100 cm^3 of water gas?

13. To 40 cm^3 of a mixture of carbon monoxide and carbon dioxide were added 30 cm^3 (an excess) of oxygen, and the mixture was sparked to initiate the following reaction

$$2CO\,(g) + O_2\,(g) \rightarrow 2CO_2\,(g)$$

A few drops of concentrated alkali were then introduced, and the carbon dioxide present dissolved. The volume of gas left was 16 cm^3. What was the volume composition of the original mixture? All volumes were measured at the same temperature and pressure.

14. 50 cm^3 of a gas mixture containing carbon monoxide, carbon dioxide and nitrogen were shaken with caustic potash and the residual volume was 36 cm^3. To this residual mixture of gases was added 20 cm^3 (an excess) of oxygen and it was exploded by sparking. After cooling, the new volume was 48 cm^3. Assuming all volumes were measured at the same room temperature and pressure, calculate the composition of the original mixture. (L)

15. The following results were obtained with a sample of the air expelled from tap water by boiling. When 50 cm^3 of the air was shaken with caustic soda solution a decrease in volume of 1.5 cm^3 was observed. Excess hydrogen was then added to the residual gas and the mixture of gases was sparked electrically. When the remaining gases had cooled they were found to occupy 45 cm^3 less than before sparking. All volumes were measured at room temperature and pressure. Calculate the percentage composition by volume of the air. (L)

16. 10 cm^3 of a hydrocarbon gas (C_xH_8) were mixed with 70 cm^3 (an excess) of oxygen (O_2) and the mixture was exploded. After allowing the resulting mixture to cool a little, concentrated alkali was introduced and the volume of the mixture decreased by 40 cm^3. All gas volumes were measured at 18°C and 755 mmHg. Calculate the value of x.

17. 20 cm^3 of a hydrocarbon of formula C_xH_4 required 60 cm^3 of oxygen for complete combustion, both volumes being measured at the same temperature and pressure. Calculate the value of x. (C)

18. A solid element X has an atomic weight of 32 and burns in excess of oxygen to form a gaseous oxide Y. Measurements carried out during this combustion show that the volume after combustion is equal to the volume before combustion provided both these volumes are measured under similar conditions. The relative (vapour) density of Y is 32.

(a) How many atoms of oxygen are there in one molecule of Y?

(b) What is the molecular weight of Y?

(c) Write the formula for Y.

19. Cyanogen is a compound of carbon and nitrogen only. 20 cm^3 of cyanogen were mixed with 60 cm^3 of oxygen (an excess) and the mixture sparked in a eudiometer tube. The two gases reacted to form carbon dioxide and nitrogen.

A little concentrated alkali was introduced into the tube and the volume then decreased by 40 cm^3. Introduction of alkaline pyrogallol (a reagent which dissolves oxygen) caused a further volume reduction and 20 cm^3 of nitrogen were left.

(a) What is the molecular formula of cyanogen?

(b) Write the equation for the reaction between cyanogen and oxygen.

All volumes were measured at the same temperature and pressure.

20. Two volumes of a gaseous hydride of antimony (Sb) yield three volumes of hydrogen on thermal decomposition. The vapour density of the hydride is 62.5. What is its formula? All volumes were measured at the same temperature and pressure.

21. 10 cm^3 of a hydrocarbon of empirical formula C_2H_5 gave 40 cm^3 of carbon dioxide when it was burned completely (all volumes measured at s.t.p.). Find its molecular formula. (O & C)

22. Find the formula of a gaseous oxide of nitrogen from the following data:

10 cm^3 of the oxide were mixed with 20 cm^3 of hydrogen and the mixture was exploded by passing a spark. When the apparatus had cooled to the original temperature the volume of the mixture of hydrogen and nitrogen was found to be 20 cm^3. 10 cm^3 of oxygen were now added, and the mixture again exploded. All the hydrogen combined with oxygen, and the mixture of unchanged nitrogen and unused oxygen occupied 15 cm^3 All measurements were made at room temperature and pressure. (L)

23. It is found that 4 g of oxygen, 7 g of the gaseous compound butene and 10.5 g of the gaseous element krypton all occupy the same volume at the same temperature and pressure.

(i) What fraction of a gramme-molecule is 4 g of oxygen?

(ii) Calculate the molecular weights of butene and krypton.

(iii) The empirical (simplest) formula of butene is CH_2. What is its molecular formula?

(iv) The atomic weight of krypton is 84. What is the atomicity of krypton?

(v) Write the equation for the burning of butene in oxygen to give carbon dioxide and water.

(vi) If 10 cm^3 of butene are burnt according to your equation, calculate the volume of oxygen used and the volume of carbon dioxide formed. (All volumes at the same temperature and pressure). (C)

7
STANDARD SOLUTIONS

1. Calculate the number of grammes of solute per dm^3 in each of the following solutions.
- (a) 0.10 M NaCl
- (b) 1.00 M $AgNO_3$
- (c) 0.05 M $CuSO_4, 5H_2O$
- (d) 2.06 M KBr
- (e) 1.02 M $NaNO_3$
- (f) 1.50 M CH_3COOCH_3

2. What is the molarity of
- (a) 0.1 M $CuSO_4$ with respect to (i) Cu^{2+} ions, (ii) $SO_4{}^{2-}$ ions?
- (b) 0.3 M $Al_2(SO_4)_3$ with respect to (i) Al^{3+} ions, (ii) $SO_4{}^{2-}$ ions?
- (c) 1.2 M H_2SO_4 with respect to (i) H^+ ions, (ii) $SO_4{}^{2-}$ ions?
- (d) 2.4 M $CaCl_2$ with respect to (i) Ca^{2+} ions, (ii) Cl^- ions?

3. Calculate the weight of the given compound required to make up a solution of the volume and molarity indicated.

Compound	Volume of solution	Molarity of solution
(a) Na_2CO_3	1 dm^3	1.0 M
(b) $CuSO_4, 5H_2O$	100 cm^3	0.5 M
(c) $Ba(OH)_2$	250 cm^3	0.01 M
(d) $AgNO_3$	2 dm^3	0.02 M
(e) $Al_2(SO_4)_3, 18H_2O$	500 cm^3	0.15 M
(f) CH_3CH_2OH	250 cm^3	0.5 M

4. The table below indicates the weights of various compounds that were used to prepare the solutions of the stated volumes. Calculate the molarities of the solutions.

Compound	Weight of compound	Volume of solution
(a) $KMnO_4$	15.8 g	1 dm^3
(b) $Na_2S_2O_3, 5H_2O$	31.0 g	250 cm^3
(c) KI	4.0 g	500 cm^3
(d) $(COOH)_2, 2H_2O$	20.0 g	100 cm^3
(e) $(NH_4)_2SO_4, FeSO_4, 6H_2O$	9.8 g	250 cm^3

5. Calculate the weight of the given compound required to make up a solution of the volume and molarity indicated. (Note that the molarity is required with respect to a given type of ion.)

Compound	Volume of solution	Molarity of solution
(a) $FeCl_3, 6H_2O$	250 cm^3	1.0 M with respect to Fe^{3+}
(b) H_2SO_4	2 dm^3	0.5 M with respect to H^+
(c) $NaBr$	100 cm^3	0.1 M with respect to Br^-
(d) $Na_2HPO_4, 12H_2O$	500 cm^3	1.5 M with respect to $HPO_4{}^{2-}$
(e) $Cr_2(SO_4)_3, 18H_2O$	250 cm^3	0.2 M with respect to Cr^{3+}

6. The molarity of concentrated hydrochloric acid is approximately 10 M. What volume of it would be required to make 500 cm^3 of approximately 0.25 M solution?

7. What is the molarity of anhydrous acetic acid (CH_3COOH)? Its density is 1.05 $g\,cm^{-3}$.

8. A solution of sodium hydroxide was titrated with 0.1 M hydrochloric acid using methyl orange as the indicator.
   25 cm^3 of the alkali required 23.2 cm^3 of the acid for neutralisation. Calculate (a) the molarity of the alkali, (b) the number of grammes of solid NaOH per dm^3.
   The reaction is represented by the equation

$$NaOH(aq) + HCl(aq) \rightarrow NaCl(aq) + H_2O(l)$$

9. A decimolar (0.1 M) solution of oxalic acid was used to determine the concentration of a solution of potassium hydroxide, the equation for the reaction being
$$H_2C_2O_4(aq) + 2KOH(aq) \rightarrow K_2C_2O_4(aq) + 2H_2O(l)$$
   25 cm^3 of oxalic acid solution required 21.5 cm^3 of the alkali for neutralisation, using phenolphthalein as the indicator.
   Calculate (a) the molarity of the potassium hydroxide solution, (b) the weight of solid potassium hydroxide in 1 dm^3 of this solution.

10. Silver nitrate ($AgNO_3$) solution reacts with any chloride solution according to the equation

$$Ag^+(aq) + Cl^-(aq) \rightarrow AgCl(s)$$

A suitable indicator (potassium chromate solution) can show when all the chloride ions have been removed from the solution as a precipitate of silver chloride.

A solution of potassium chloride (KCl) was titrated with 0.1 M silver nitrate with the following result:

$$25 \text{ cm}^3 \text{ KCl solution} \equiv 26.4 \text{ cm}^3 \text{ AgNO}_3 \text{ solution}$$

Calculate the molarity of the potassium chloride solution and the number of grammes of solid KCl in 1 dm^3 of the solution.

11. An acid reacts with an alkali according to the equation

$$H^+(aq) + OH^-(aq) \rightarrow H_2O(l)$$

A solution of a monobasic acid (i.e. one molecule of it yields one hydrogen ion in solution) of molarity 0.1 M was titrated against an alkaline solution, using a suitable indicator. 25 cm^3 of the alkali required 21.2 cm^3 of the acid for neutralisation.

Calculate (a) the molarity of the alkaline solution with respect to OH^- ions, (b) the weight in grammes of OH^- ions per dm^3.

12. 5.60 g of an acid of molecular weight 224 were dissolved in water and made up to 250 cm^3. 10 cm^3 of this solution, when titrated with 0.1 M potassium hydroxide solution, required 40.2 cm^3 of the alkali for neutralisation. What was the molarity of the acid solution? How many g-ions of H^+ did 250 cm^3 of the acid solution contain?

13. 7.15 g of hydrated sodium carbonate (Na_2CO_3, xH_2O) were weighed out, dissolved in water and the solution was made up to 250 cm^3. 25 cm^3 of this solution were titrated with 0.227 M hydrochloric acid. It was found that 22.0 cm^3 of acid were required. Calculate the value of x.

The equation for the reaction is
$$Na_2CO_3(aq) + 2HCl(aq) \rightarrow 2NaCl(aq) + CO_2(g) + H_2O(l)$$

14. Sulphamic acid has the formula H_3NSO_3 and is monobasic; it is a strong acid and reacts with alkali carbonates quantitatively, using methyl orange as indicator. It was found that 30 cm^3 of a solution containing 24.25 g dm^{-3} of this acid neutralised 25 cm^3 of a solution of sodium carbonate. Calculate the concentration of the alkali in grammes Na_2CO_3 per dm^3. (O)

15. When a solution of sulphuric acid was standardised the following results were obtained: exactly 5.3 g of pure anhydrous sodium carbonate were dissolved in water and made up to 250 cm^3 of solution; 25 cm^3 of this solution were neutralised by 20 cm^3 of the acid.

Calculate what volumes of this acid would be required (a) exactly to dissolve 1.92 g of magnesium, and (b) to precipitate 58.25 g barium sulphate from a solution of barium chloride.　　　　　　　　　　　　　　　　　　　　　　　　　　　　　(O)

16. 25 cm^3 of a solution of sodium hydroxide were placed in a polystyrene beaker, together with a thermometer. The temperature was noted and the 1.0 M nitric acid was added, 5 cm^3 at a time, the temperature being noted after each addition. The temperature readings were plotted against 'volume of acid added', and the following graph was obtained

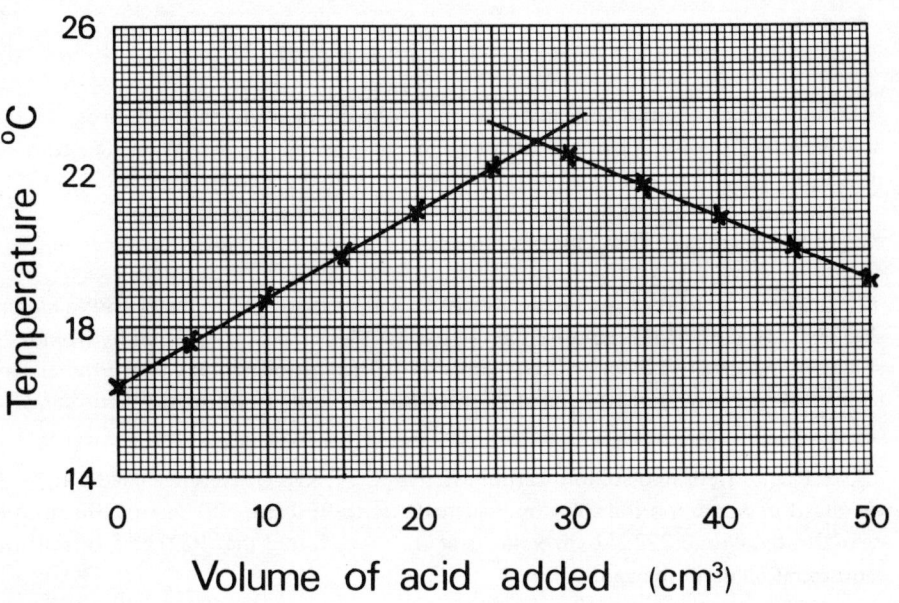

The equation for the reaction is
$$NaOH(aq) + HNO_3(aq) \rightarrow NaNO_3(aq) + H_2O(l)$$

(a) What volume of acid was required to neutralise this 25 cm^3 of alkali?
(b) What was the molarity of the alkali?
(c) What weight of solid sodium hydroxide (NaOH) was contained in 1.0 dm^3 of the solution?
(d) What was the molarity of the final, mixed, solution with respect to Na$^+$ ions?

17. A series of volumes of aqueous potassium hydroxide were treated with volumes of 1.5 M hydrochloric acid, the total volume of each mixture always being 20 cm^3. The temperature rise was measured in each case and the following table summarises the results:

Vol. KOH(aq) cm^3	2	4	6	8	10	12	14	16	18
Vol. 1.5 M HCl cm^3	18	16	14	12	10	8	6	4	2
Temp. rise $^\circ$C	1.9	3.8	5.7	7.6	7.5	6.0	4.5	3.0	1.5

The equation representing the reaction is
$$KOH(aq) + HCl(aq) \rightarrow KCl(aq) + H_2O(l)$$

Plot a graph of these results and use this to determine the molarity of the potassium hydroxide solution. What is the concentration in g dm^{-3} of the potassium hydroxide (KOH)?

18. 25 cm^3 of a 0.1 M solution of barium hydroxide were placed in a titration flask that was fitted with electrodes. The conductance of the solution (its ability to carry an electric current) was measured. Aqueous sulphuric acid was then added, 5.0 cm^3 at a time, the conductance being measured after each addition. A precipitate formed during the titration, the reaction being represented by the equation
$$Ba(OH)_2(aq) + H_2SO_4(aq) \rightarrow BaSO_4(s) + 2H_2O(l)$$
A graph of the results is shown below

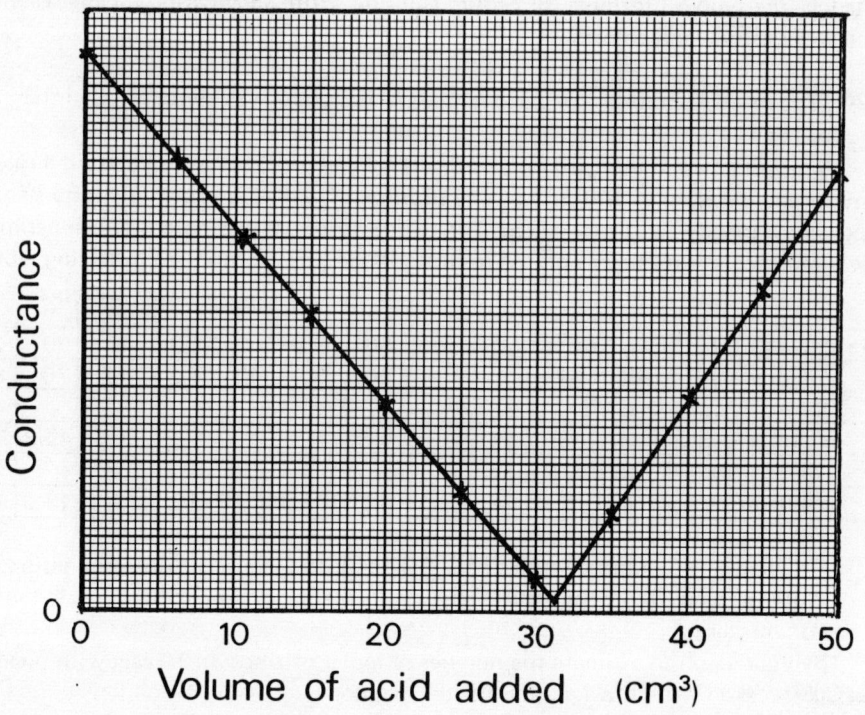

Account for the shape of the graph, including the decrease of the conductance almost to zero at one point. The ionic equation is

$$Ba^{2+}(aq) + 2OH^-(aq) + 2H^+(aq) + SO_4{}^{2-}(aq) \rightarrow BaSO_4(s) + 2H_2O(l)$$

What is the molarity of the sulphuric acid solution?

19. 14 g of cerium (Ce), a metal of atomic weight 140, were converted to a chloride of formula $CeCl_x$ by heating in a stream of chlorine (Cl_2). All of the prepared chloride was dissolved in water and made up to 100 cm^3. 1 cm^3 of this solution was placed in each of six test-tubes of equal bore. 1.0 M silver nitrate solution and water were added to each of the test-tubes in the quantities indicated in the table.

Test-tube	1	2	3	4	5	6
Vol. 1.0 M AgNO$_3$ (cm^3)	1	2	3	4	5	6
Vol. CeCl$_x$ soln. (cm^3)	1	1	1	1	1	1
Vol. water (cm^3)	8	7	6	5	4	3
Height of precipitate (cm)	0.4	0.8	1.2	1.6	1.6	1.6

Silver chloride was precipitated, the equation for the reaction being

$$Ag^+(aq) + Cl^-(aq) \rightarrow AgCl(s)$$

After settling, the height of each precipitate was measured and recorded as above.

Plot a graph of 'height of precipitate' against 'volume of 1.0 M AgNO$_3$ solution' and deduce the simplest formula for cerium chloride. Write an equation for the reaction between cerium and chlorine.

20. The reaction between zinc and copper (II) sulphate was investigated in the following way.

A measured volume of 1.0 M copper (II) sulphate solution was placed in a calorimeter and made up to 100 cm^3 by the addition of water. A known quantity of zinc powder was added and well mixed. The temperature rise was noted. The experiment was repeated several times and the results obtained are given in the following table:

	1	2	3	4	5	6	7
Vol. of 1.0 M CuSO$_4$ (cm^3)	90	70	60	50	40	30	20
Vol. of water added (cm^3)	10	30	40	50	60	70	80
Moles of CuSO$_4$ present							
Wt. of zinc added (g)	0.65	1.96	2.62	3.27	3.92	4.58	5.23
Moles of zinc (Zn) added							
Temperature rise (°C)	4.3	11.6	15.8	19.1	19.1	14.6	10.0

(i) Calculate the number of moles of CuSO$_4$ and Zn used in each determination.

(ii) Plot a graph of 'temperature rise' against 'percentage composition' (mole/mole) of reaction mixture.

Use your graph to estimate the number of moles of zinc which react with one mole of CuSO$_4$; then write the equation for the reaction.

8
THERMOCHEMISTRY

Heats (or enthalpies) of reaction, and heat capacities, in this section are given both in kJ and kcal. In your calculations use <u>either</u> the kJ values <u>or</u> the kcal values. Be careful not to mix them up.

The heat capacity of water and of aqueous solutions may be taken as 4.18 kJ (1 kcal) $kg^{-1} \, {}^{\circ}C^{-1}$.

1. The following equation indicates that when 1 mole of carbon (C) burns completely in air or oxygen 393 kJ (94 kcal) of heat are evolved.

$$C(s) + O_2 (g) \rightarrow CO_2 (g); \Delta H = -393 \text{ kJ } (-94 \text{ kcal}) \text{ g}-\text{eq}^{-1}$$

What quantity of heat would be given out on complete combustion of

(a) 10 moles of carbon

(b) 0.25 mole of carbon

(c) 12 g of carbon

(d) 18 g of carbon?

What weight of carbon would have to be burned to produce

(e) 196.5 kJ (47 kcal) of heat

(f) 786 kJ (188 kcal) of heat

(g) sufficient heat to raise the temperature of 2 dm^3 of water from 25°C to 90°C?

2. When calcium carbonate is precipitated by adding a solution of sodium carbonate to one of calcium chloride, the change may be represented by

$$Ca^{2+}(aq) + CO_3{}^{2-}(aq) \rightarrow CaCO_3 (s); \Delta H = + 12.6 \text{ kJ}(+3.0 \text{ kcal}) \text{ g}-\text{eq}^{-1}$$

(a) Is the reaction exothermic or endothermic?

What would be the heat (enthalpy) change if

(b) 0.5 mole of calcium chloride (Ca^{2+}, $2Cl^-$) was treated, in aqueous solution, with a slight excess of aqueous sodium carbonate ($2Na^+$, $CO_3{}^{2-}$),

(c) 111 g of calcium chloride were similarly treated,

(d) 1 mole of calcium nitrate (Ca^{2+}, $2NO_3{}^-$) and 1 mole of potassium carbonate ($2K^+$, $CO_3{}^{2-}$) were reacted in aqueous solutions?

3. When 6.55 g of zinc were dissolved in dilute sulphuric acid 15.5 kJ (3.7 kcal) of heat were evolved. Complete the following thermochemical equation (remember that such an expression represents molar quantities):

$$Zn(s) + 2H^+(aq) \rightarrow Zn^{2+}(aq) + H_2(g); \Delta H = \ldots\ldots$$

4. The thermochemical equation for the reaction between zinc and dilute hydrochloric acid (HCl) may be written

$$Zn(s) + 2H^+(aq) \rightarrow Zn^{2+}(aq) + H_2(g); \Delta H = -150 \text{ kJ} (-36 \text{ kcal}) \text{ g-eq}^{-1}$$

(a) Is the reaction exothermic or endothermic?
(b) How many moles of (i) zinc (Zn), (ii) hydrochloric acid (HCl), react when the heat (enthalpy) change is -150 kJ (-36 kcal)?
(c) What would be the heat (enthalpy) change if 2.0 moles of zinc were dissolved in excess dilute hydrochloric acid?
(d) If the heat (enthalpy) change were -112.5 kJ (-27 kcal)
(i) What weight of zinc would have dissolved in excess dilute hydrochloric acid?
(ii) What weight of hydrogen would have been produced?
(e) If dilute sulphuric acid were used instead of hydrochloric acid how many moles of this acid (H_2SO_4) would react if the heat (enthalpy) change was -150 kJ (-36 kcal), excess zinc being present?

5. If 6.0 g of carbon burn completely in oxygen, forming carbon dioxide with the liberation of 196.6 kJ (47 kcal) of heat, what is the heat (enthalpy) of combustion of 1 g-atom of carbon? Write the thermochemical equation for this heat (enthalpy) of combustion.

6. 4.0 g of methane (CH_4) burn completely in air to form carbon dioxide and water, with the evolution of 334 kJ (80 kcal) of heat. What is the heat (enthalpy) of combustion of 1 mole of methane? Write the thermochemical equation.

7. In each of the following examples complete the thermochemical equation:
(a) When 1.15 g of sodium reacted with excess water 90.7 kJ (21.1 kcal) of heat were evolved.

$$Na(s) + H_2O(l) \rightarrow NaOH(aq) + \tfrac{1}{2}H_2(g); \Delta H = \ldots\ldots$$
$$2Na(s) + 2H_2O(l) \rightarrow 2NaOH(aq) + H_2(g); \Delta H = \ldots\ldots$$

(b) When 1.0 g of the gas ethane was burned in excess oxygen 50 kJ (12 kcal) of heat were evolved.

$$2C_2H_6(g) + 7O_2(g) \rightarrow 4CO_2(g) + 6H_2O(g); \Delta H = \ldots\ldots$$

How much heat is evolved when 1 mole of ethane is burnt in excess oxygen?

(c) When 4.9 g of concentrated sulphuric acid were diluted with sufficient water to make a 1.0 M solution 4.18 kJ (1.0 kcal) of heat were evolved.

$$H_2SO_4(l) + aq \rightarrow H_2SO_4(aq, 1.0 \text{ M}); \Delta H = \ldots\ldots$$

(d) When 24.95 g of copper (II) sulphate crystals ($CuSO_4$, $5H_2O$) were dissolved in a large quantity of water 11.5 kJ (2.75 kcal) of heat were absorbed.

$$CuSO_4, 5H_2O(s) + aq \rightarrow CuSO_4(aq); \quad \Delta H = ...$$

8. From the following energy diagram deduce the heat (enthalpy) change which would occur if 10.0 g of ammonium chloride were dissolved in a large quantity of water.

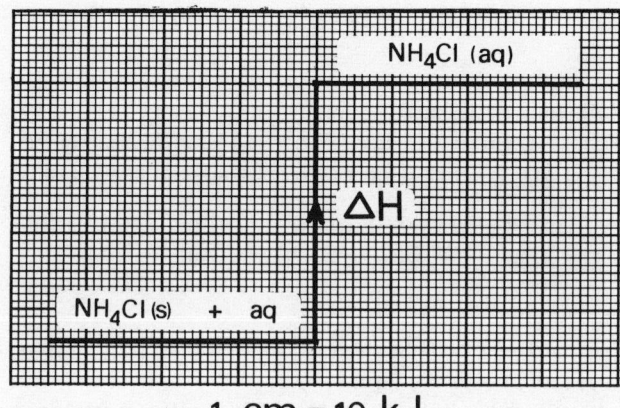

$$1 \ cm \equiv 10 \ kJ$$

9. The heat (enthalpy) change which occurs when an acid and an alkali react in aqueous solution may be represented by the equation

$$H^+(aq) + OH^-(aq) \rightarrow H_2O(l); \quad \Delta H = -57 \ kJ \ (-13.6 \ kcal) \ g\!-\!eq^{-1}$$

It may also be represented by an energy diagram of the type

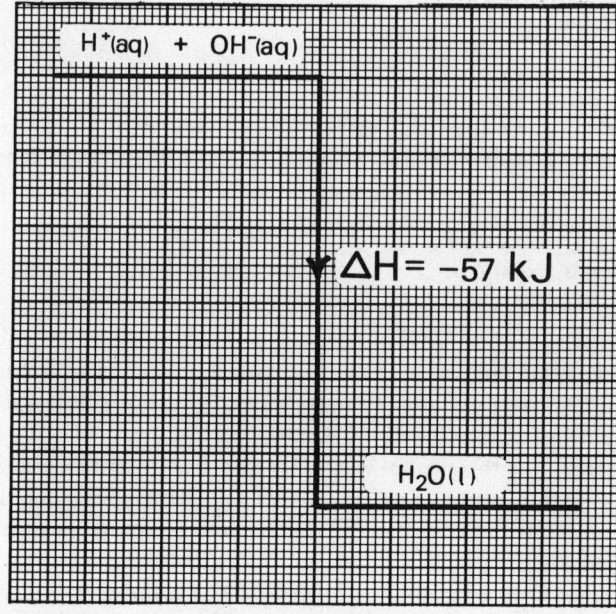

Draw energy diagrams, to scale, on graph paper representing the heat (enthalpy) changes which occur in the following reactions.

(a) $C(s) + 2S(s) \rightarrow CS_2$ (l); $\Delta H = +121$ kJ (+29.0 kcal) g–eq^{-1}
(b) H_2 (g) $+\frac{1}{2}O_2$ (g) $\rightarrow H_2O$(l); $\Delta H = -266$ kJ (−63.5 kcal) g–eq^{-1}
(c) NH_4NO_3 (s) + aq $\rightarrow NH_4NO_3$ (aq); $\Delta H = +26$ kJ (+6.2 kcal) g–eq^{-1}
(d) $Zn(s) + 2H^+$(aq) $\rightarrow Zn^{2+}$(aq) $+ H_2$ (g); $\Delta H = -150$ kJ (−36 kcal) g–eq^{-1}
(e) $P(s) + 1\frac{1}{2}Cl_2$ (g) $\rightarrow PCl_3$ (l); $\Delta H = -450$ kJ (−108 kcal) g–eq^{-1}

10. Study the following equations:
$$C(s) + O_2 \text{ (g)} \rightarrow CO_2 \text{ (g)}; \Delta H = -393 \text{ kJ } (-94 \text{ kcal) g–eq}^{-1}$$
$$CO(g) + \tfrac{1}{2}O_2 \text{ (g)} \rightarrow CO_2 \text{ (g)}; \Delta H = -284 \text{ kJ}(-68 \text{ kcal) g–eq}^{-1}$$

What would you expect the value of ΔH to be for the following combustion?
$$C(s) + \tfrac{1}{2}O_2 \text{ (g)} \rightarrow CO(g)$$

11. The heats (enthalpies) of combustion of the two solid polymorphs (allotropes) red phosphorus and yellow phosphorus are:
$$P_4 \text{ (red)} + 5O_2(g) \rightarrow P_4O_{10} \text{ (s)}; \Delta H = -2912 \text{ kJ}(-696 \text{ kcal) g–eq}^{-1}$$
$$P_4 \text{ (yel)} + 5O_2(g) \rightarrow P_4O_{10} \text{ (s)}; \Delta H = -2929 \text{ kJ}(-700 \text{ kcal) g–eq}^{-1}$$

(a) Calculate the heat (enthalpy) change for
$$P_4 \text{ (red)} \rightarrow P_4 \text{ (yel)}$$

(b) Is the change exothermic or endothermic?
(c) Which form of phosphorus has the most energy?
(d) Which is the stable form?

12. 8.0 g of mothballs (naphthalene, $C_{10}H_8$) burn completely in oxygen with the liberation of 322 kJ (77 kcal) of heat. Complete the following thermochemical equation:
$$C_{10}H_8 \text{ (s)} + aO_2 \text{ (g)} \rightarrow bCO_2 \text{ (g)} + cH_2O(g); \Delta H = \text{......}$$

i.e. give values for a, b, c and ΔH.

13. A heat of combustion apparatus (fuel calorimeter) of water equivalent 200 g and containing 350 cm^3 of water was used to determine the heat of combustion of propan–1–ol (an alcohol of formula $CH_3CH_2CH_2OH$).

A spirit lamp containing the alcohol was weighed (12.12 g) and then lit under the apparatus, the initial temperature of which was 18.1°C. After some time the lamp was extinguished and reweighed (11.62 g) and the maximum temperature reached by the apparatus noted (25.4°C).

Calculate the molar heat (enthalpy) of combustion of propan–1–ol.

14. The heat (enthalpy) of combustion of butan—1—ol $(CH_3 CH_2 CH_2 CH_2 OH)$ was determined, using a heat of combustion apparatus of water equivalent 250 g and containing 400 g of water. A spirit lamp containing the alcohol was weighed (14.29 g) and then lit under the apparatus. After some time the lamp was extinguished and a temperature rise of $6.4°C$ in the apparatus was noted. The spirit lamp was reweighed (13.81 g).

   Calculate the molar heat (enthalpy) of combustion of butan—1—ol.

15. In a determination of the heat (enthalpy) of combustion of sulphur in oxygen (forming sulphur dioxide) the experimental results were:

Weight of crucible + sulphur before combustion	$= 12.92$ g
Weight of crucible + sulphur after combustion	$= 10.62$ g
Initial temperature of calorimeter and water	$= 15.2°C$
Final temperature of calorimeter and water	$= 22.2°C$
Thermal capacity of calorimeter	$= 1046$ J (250 cal) $°C^{-1}$
Weight of water in calorimeter	$= 650$ g

(a) How much heat was evolved during the combustion of this weight of sulphur?
(b) What is the heat (enthalpy) of combustion of sulphur (S) per g-atom?
(c) Write the thermochemical equation for the combustion of 1 g-atom of sulphur.
(d) Draw an energy diagram for this change.

16. The following experimental results were obtained for a determination of the heat (enthalpy) of combustion of carbon (graphite) in oxygen:

Weight of crucible + carbon before combustion	$= 16.31$ g
Weight of crucible + carbon after combustion	$= 15.74$ g
Weight of water in calorimeter	$= 450$ g
Water equivalent of calorimeter	$= 200$ g
Initial temperature of calorimeter and water	$= 16.3°C$
Final temperature of calorimeter and water	$= 23.2°C$

(a) What quantity of heat was evolved during the combustion of this weight of graphite?
(b) What is the heat (enthalpy) of combustion of 1 g-atom of carbon (C)?
(c) Write the thermochemical equation for the combustion of 1 g-atom of carbon.
(d) Draw an energy diagram to represent this change.

17. When 0.5 g-formula of ammonium nitrate $(NH_4 NO_3)$ is added to 475 cm^3 of water 500 cm^3 of ammonium nitrate solution are produced and the temperature falls by $6.2°C$.

   Calculate the heat (enthalpy) of solution, ΔH, in the following change
$$NH_4 NO_3 (s) + aq \rightarrow NH_4 NO_3 (aq, 1.0 \text{ M}); \Delta H = ...$$

Draw an energy diagram for this change.

18. 27 cm^3 of concentrated sulphuric acid (H_2SO_4) is very nearly 0.5 mole. When this quantity of acid was added to 475 cm^3 of water a temperature rise of $16°C$ occurred. Assuming the final volume of the mixture was 500 cm^3, complete the following thermo-chemical equation

$$H_2SO_4 (l) + aq \rightarrow H_2SO_4 (aq, ...M); \Delta H = ...$$

19. 50 cm^3 of 0.05 M silver nitrate solution were placed in a plastic calorimeter and the temperature noted. A slight excess of magnesium powder was then added and the mixture stirred well. The temperature of the solution rose by $6.8°C$. Calculate the heat (enthalpy) of the reaction

$$2Ag^+(aq) + Mg(s) \rightarrow Mg^{2+}(aq) + 2Ag(s)$$

You may assume that the thermal capacity of the calorimeter is negligible.

20. 100 cm^3 of 1.0 M copper (II) sulphate solution were treated with a slight excess of zinc powder. A temperature rise of $5.2°C$ was recorded. Calculate the heat (enthalpy) of the reaction

$$Cu^{2+}(aq) + Zn(s) \rightarrow Zn^{2+}(aq) + Cu(s)$$

Draw an energy diagram for the reaction.

21. 25 cm^3 of 1.0 M barium chloride solution and 25 cm^3 of 1.0 M sodium sulphate solution, both at $18.2°C$, were mixed in a thin plastic calorimeter of negligible heat capacity. Barium sulphate was precipitated and heat was evolved, causing the temperature to rise to $20.5°C$.

Calculate the heat (enthalpy) of the reaction

$$Ba^{2+}(aq) + SO_4{}^{2-}(aq) \rightarrow BaSO_4 (s)$$

If 50 cm^3 of each of these molar solutions were mixed what would be the temperature rise?

22. 50 cm^3 of 0.5 M silver nitrate solution at $18.0°C$ were mixed with 50 cm^3 of 0.5 M sodium bromide solution at the same temperature. Silver bromide was precipitated and the temperature rose to $23.0°C$. Calculate the heat (enthalpy) of precipitation of silver bromide

$$Ag^+(aq) + Br^-(aq) \rightarrow AgBr(s)$$

What temperature rise would you expect if the following solutions were mixed?
(a) 50 cm^3 0.5 M silver nitrate and 100 cm^3 0.5 M sodium bromide.
(b) 40 cm^3 0.5 M silver nitrate and 110 cm^3 0.5 M sodium bromide.
(c) 50 cm^3 0.5 M silver nitrate and 50 cm^3 0.5 M potassium bromide.
(d) 100 cm^3 0.5 M silver nitrate and 50 cm^3 0.5 M sodium bromide.
Heat changes due to dilution may be neglected.

23. When 50 cm^3 of 1.0 M hydrochloric acid were added to 50 cm^3 of 1.0 M sodium hydroxide, in a plastic calorimeter of negligible heat capacity, a temperature rise of 6.8°C was observed. Calculate the heat (enthalpy) of neutralisation of sodium hydroxide by hydrochloric acid and express this in terms of (a) a thermochemical equation, (b) an energy diagram.

24. The following table gives the molar heats (enthalpies) of combustion of a series of alkanes (paraffins).

Alkane	Formula	Heat (enthalpy) of combustion kJ (kcal) mol^{-1}
methane	CH_4	−890 (−213)
ethane	CH_3CH_3	−1560 (−373)
propane	$CH_3CH_2CH_3$	−2220 (−531)
butane	$CH_3(CH_2)_2CH_3$	−2880 (−688)
pentane	$CH_3(CH_2)_3CH_3$	−3510 (−840)
hexane	$CH_3(CH_2)_4CH_3$	−4160 (−995)

(a) Plot a graph of 'heat (enthalpy) of combustion' against 'number of carbon atoms per molecule'.
(b) What would you expect the heat (enthalpy) of combustion of the next member of the series, heptane $CH_3(CH_2)_5CH_3$, to be?
(c) Another alkane has a heat (enthalpy) of combustion of −6780 kJ (−1620 kcal) mol^{-1}. How many carbon atoms has this alkane per molecule? Write its formula.

25. Ammonia reacts with hydrogen chloride to form ammonium chloride and the reaction can be represented by the following energy level diagram.

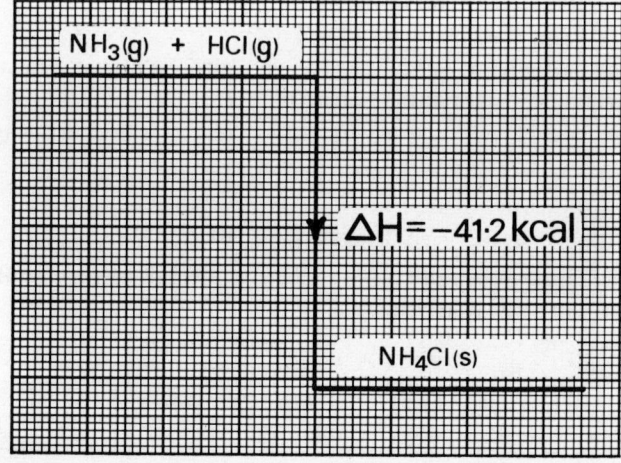

NH$_3$(g) + HCl(g)

$\Delta H = -41\cdot2$ kcal

NH$_4$Cl(s)

(a) Suppose 1 g-formula of ammonia was mixed with 2 g-formulae of hydrogen chloride. Calculate the energy change which would take place. State whether the change would be exothermic or endothermic.

(b) When 1 g-formula of ammonia is dissolved in water to form a dilute solution 8.4 kcal are given out. Draw an energy level diagram (on the same scale as the one above) to represent the change.

(c) When 1 g-formula of hydrogen chloride is dissolved in water to form a dilute solution, 17.3 kcal are given out. When the solutions of ammonia and hydrogen chloride are mixed to form a solution of ammonium chloride, 12.5 kcal are given out. Calculate ΔH for the reaction:

$$NH_3(g) + HCl(g) + aq \rightarrow NH_4Cl(aq)$$

(L, Nuff.)

Molar heats of vaporisation

26. 100 g of tetrachloromethane (carbon tetrachloride) were heated in a flask, in a fume cupboard, by means of a small immersion heater. In 3 minutes the temperature rose by $40°C$. The heat capacity of tetrachloromethane is 0.88 kJ (0.21 kcal) $°C^{-1}$ kg^{-1}.

(a) How much heat was supplied by the immersion heater in 3 minutes?

(b) At what rate was the immersion heater supplying heat?

The heat capacity of the flask may be ignored.

27. 100 g of water in a conical flask were heated by a bunsen flame. During the first 5 minutes the temperature rose by $50°C$. Without altering the flame, the water was allowed to come to the boil, and boiling was continued for 10 minutes. When boiling was stopped, 80 g of water were found to remain.

(a) At what rate was the bunsen supplying heat?

(b) How much heat energy was required to convert 20 g of water to steam while the temperature remained constant at the boiling point?

(c) How much energy would be required to convert 1 mole of water to steam?

28. 150 g of ethanol (CH_3CH_2OH) were heated in a flask by means of a small immersion heater. The temperature was noted after intervals of time until it had risen to about $65°C$; the results are given below.

Time (min)	0	0.5	1.0	1.5	2.0	2.5	3.0	3.5	4.0
Temp. ($°C$)	15	22	27.5	35	39.5	47	53	61	66.5

Heating was continued and the ethanol allowed to boil. Boiling was continued for 10.0 minutes. On cooling, the ethanol had lost 51.5 g in weight.

(a) Use a graphical method to determine the rate at which the immersion heater was supplying energy. The heat capacity of liquid ethanol is 111 J (26.6 cal) $°C^{-1}$ mol^{-1}.

(b) How much energy was used up in boiling off the ethanol?

(c) How much energy would be required to convert 1 mole of liquid ethanol to the vapour? (i.e. calculate the molar heat (enthalpy) of vaporisation of ethanol).

29. Each of six liquids was brought to the boil, in turn, by the same immersion heater, and boiling was continued for a given period. The output of the heater was 6.28 kJ (1.5 kcal) min^{-1}. The following table indicates the time for which each liquid was boiled, and the weight of the liquid before and after the experiment.

Liquid	Formula	Boiling time (min)	Wt. before (g)	Wt. after (g)
(a) dichloromethane	CH_2Cl_2	3.0	160.0	102.0
(b) trichloromethane	$CHCl_3$	3.5	190.0	100.4
(c) tetrachloromethane	CCl_4	3.0	255.0	160.1
(d) dibromomethane	CH_2Br_2	6.0	400.0	204.3
(e) tribromethane	$CHBr_3$	5.5	440.0	207.5
(f) tetrabromomethane	CBr_4	4.0	450.0	246.7

(i) Calculate the molar heat (enthalpy) of vaporisation of each liquid.

(ii) Plot a graph of molar heat (enthalpy) of vaporisation against molecular weight for these compounds and comment on the curve.

9
ELECTROCHEMISTRY

1. How many Faradays of electricity would be required to liberate the following quantities of elements:
(a) 1 g-atom of silver from a solution containing Ag^+ ions,
(b) 63.5 g of copper from a solution containing Cu^{2+} ions,
(c) 2 g-atoms of lead from a solution containing Pb^{2+} ions,
(d) 1 mole of chlorine (Cl_2) from a solution containing Cl^- ions,
(e) 23 g of sodium from fused sodium chloride which contains Na^+ ions,
(f) 11.2 dm^3 of hydrogen (H_2) at s.t.p. from a solution of an acid (which contains H^+ ions)?

2. A certain quantity of electricity caused the deposition of 0.1 g-atom of silver from a silver nitrate solution by the discharge of Ag^+ ions.

How many moles of the following elements would be produced by the same quantity of electricity by the discharge of the following ions:
(a) zinc by the discharge of Zn^{2+} ions,
(b) aluminium by the discharge of Al^{3+} ions,
(c) sodium by the discharge of Na^+ ions,
(d) hydrogen (H_2) by the discharge of H^+ ions,
(e) mercury (Hg) by the discharge of Hg^{2+} ions,
(f) mercury (Hg) by the discharge of Hg_2^{2+} ions?

3. Calculate the charge on each of the ions discharged in the following electrolyses and so write the formula of the ion.
(a) 0.59 g of nickel (Ni) was deposited at the cathode, from a solution containing nickel ions, by a current of 2 A flowing for 16 min.
(b) 1.08 g of silver was deposited at the cathode, from a silver nitrate solution, by a current of 3 A flowing for 5 min 22 sec.
(c) 1.3 g of bromine was liberated at the anode when a solution of bromide ions was electrolysed using a current of 4 A which flowed for 6.5 min.
(d) 10.35 g of lead was liberated at the cathode from molten lead bromide by a current of 1 A flowing for 2 h 40 min.

4. If 1 g-atom of a metallic element M requires three Faradays of charge for electrolytic deposition, give the formulae for its oxide, chloride and sulphate. (O)

5. (a) How many moles of hydrogen (H_2) would be produced by the passage of 4 Faradays of electricity through a solution of H^+ ions?
(b) How many moles of silver would be deposited by a current of 2 A flowing for 10 min through a solution containing Ag^+ ions?
(c) How many grammes of lead would be deposited on the cathode, from an electrolyte containing Pb^{2+} ions by 0.2 Faraday of electricity?
(d) How many moles of iodine (I_2) would be liberated during the electrolysis of a potassium iodide (K^+, I^-) solution using a current of 1 A for 20 min?

6. If the same current is passed through solutions of silver nitrate $(AgNO_3)$ and copper (II) sulphate $(CuSO_4)$, how much silver will be liberated in the same time as 1.00 g of copper? (L)

7. The same quantity of current was passed through three voltameters, depositing copper in the first, silver in the second and liberating 200 cm^3 of hydrogen (at s.t.p.) in the third. Calculate the weights of copper and silver deposited. (L)

8. What *total* volume of gas, measured at s.t.p., would be liberated from dilute sulphuric acid by the current which, flowing for the same time, deposits 0.3175 g of copper from copper (II) sulphate solution? (L)

9. Calcium metal may be obtained by the electrolysis of fused calcium chloride $(Ca^{2+}, 2Cl^-)$. How long (to the nearest minute) would a current of 2 A have to be passed in order to obtain 2 g of calcium?

10. Copper (II) sulphate solution was electrolysed for 30 min using a steady current. The cathode was weighed before and after electrolysis and the increase in weight was found to be 0.623 g. What was the current?

11. What current flowing for 5 min will liberate 100 cm^3 of hydrogen (H_2) from a solution of H^+ ions if the gas is measured at $20°C$ and 780 mmHg pressure?

12. When an aqueous solution of sodium hydroxide is electrolysed using platinum electrodes the hydroxyl ions are discharged at the anode to give oxygen according to the following overall equation:

$$4OH^-(aq) \rightarrow 2H_2O(l) + O_2(g) + 4e$$

If a current of 2 A is passed for 15 min
(a) How many moles of OH^- ions are discharged?
(b) How many moles of oxygen (O_2) are produced?
(c) What is the volume of this oxygen expressed at s.t.p?

13. A steady current of 1 A is passed between copper electrodes through a solution of copper (II) sulphate for 1 hour. The cathode gains in weight by 1.19 g and the anode loses in weight by the same amount.

In another experiment, a steady current of 2 A is passed for 1 hour between copper electrodes in a warm alkaline solution of sodium chloride. The anode is found to lose in weight by 4.76 g.

Calculate the number of Faradays of electricity required to cause 1 g-atom of copper (63.6 g) to dissolve from the anode in each case, and comment on the answer. (L)

14. An aqueous solution of sodium acetate (Na^+, CH_3COO^-) was electrolysed. The reactions which occurred at the electrodes may be represented by the following equations:

cathode $2H^+(aq) + 2e \rightarrow H_2(g)$
anode $2CH_3COO^-(aq) \rightarrow C_2H_6(g) + 2CO_2(g) + 2e$

(a) What were the relative volumes of the gases collected at the cathode and anode respectively? (The solubility of carbon dioxide in water may be neglected).
(b) After passing the anode gas through sodium hydroxide solution the ethane (C_2H_6) was collected and its volume measured. This was found, after correction to s.t.p., to be 25 cm^3. How many Faradays of electricity had been passed through the sodium acetate solution?
(c) What volume of (i) hydrogen, (ii) carbon dioxide, expressed at s.t.p., were also produced during this electrolysis?

15. A solution of potassium auricyanide was electrolysed, using a gold anode and a gold cathode. During the electrolysis gold was deposited on the cathode. The electrodes were weighed before and after the experiment.

	cathode	anode
Initial weight	25.104 g	24.614 g
Final weight	25.596 g	24.122 g
Current flowing	0.20 A	
Time of current flow	60 min	

(a) Calculate the amount of electricity which flowed during the experiment. Give your answer in amp-hours or coulombs.
(b) What weight of gold would have been deposited by 1 Faraday of electricity?
(1 Faraday = 26.75 amp-hours or 96 500 coulombs)
(c) The atomic weight of gold is 197. How many Faradays of electricity are required to deposit 1 g-atom of gold?
(d) What does this experiment tell us about the number and kind of charges on one gold atom? (L, Nuff.)

16. 50 cm^3 of a solution of iron (III) chloride containing 0.1 g-ion of Fe^{3+} per litre were heated till nearly boiling and then a solution of tin (II) chloride B (containing 1 g-ion of Sn^{2+} per litre) was added slowly; when 2.5 cm^3 of the latter had been added the yellow colour of the iron (III) solution had just been removed. At this stage it may be assumed that all the iron (III) chloride had been converted to iron (II) and all the tin (II) had been converted to an ion with a higher formal charge.

(a) How many g-ions of iron (III) were present in the 50 cm^3 of solution A?

(b) How many g-ions of tin (II) were present in 2.5 cm^3 of solution B?

(c) How many ions of Fe^{3+} have reacted with one ion of Sn^{2+}?

(d) What is the formula of the tin ion which has been formed?

(e) Write an equation for the reaction.

(f) If some of solution B is electrolysed with a current of 1 amp for 160 min (i.e. 2.67 h), how many g-atoms of tin would you expect to be deposited at the cathode?

(L, Nuff.)

Cells and standard electrode potentials

The data in the following table may be used where necessary in answering questions 17 – 20

Standard electrode potentials at 25°C			$E^{\ominus}$ (Volts)
	Zn^{2+}(aq)	Zn(s)	− 0.76
	Pb^{2+}(aq)	Pb(s)	− 0.13
Pt, H_2 (g) | H^+(aq)	H^+(aq)	H_2 (g), Pt	0.00
	Cu^{2+}(aq)	Cu(s)	+ 0.34
	Ag^+(aq)	Ag(s)	+ 0.80

17. Calculate the e.m.fs' of the following cells, at 25°C, and state which metal forms the positive terminal of the cell.

(a)	Pt, H_2 (g)	I	H^+(aq)	Zn^{2+}(aq)	I	Zn(s)
(b)	Zn(s)	I	Zn^{2+}(aq)	Cu^{2+}(aq)	I	Cu(s)
(c)	Ag(s)	I	Ag^+(aq)	Pb^{2+}(aq)	I	Pb(s)
(d)	Pt, H_2 (g)	I	H^+(aq)	Cu^{2+}(aq)	I	Cu(s)
(e)	Ag(s)	I	Ag^+(aq)	Cu^{2+}(aq)	I	Cu(s)
(f)	Zn(s)	I	Zn^{2+}(aq)	Ag^+(aq)	I	Ag(s)

18. For each of the cells in the previous question write:

(i) the half-equation for the reaction at the positive electrode,

(ii) the half-equation for the reaction at the negative electrode,

(iii) the overall equation for the cell by combining the half-equations.

19. Calculate the standard electrode potentials of the underlined electrode systems in the following cells of given e.m.fs' at $25°C$ (assume standard conditions):

(a)	Pt, H_2 (g)	$\mid$	H^+(aq)	$\vdots$	Hg^{2+}(aq)	$\mid$	Hg(l)	E = + 0.79 V
(b)	Mg(s)	$\mid$	Mg^{2+}(aq)	$\vdots$	Cu^{2+}(aq)	$\mid$	Cu(s)	E = + 2.71 V
(c)	Sn(s)	$\mid$	Sn^{2+}(aq)	$\vdots$	Ag^+(aq)	$\mid$	Ag(s)	E = + 0.94 V
(d)	Au(s)	$\mid$	Au^{3+}(aq)	$\vdots$	Ag^+(aq)	$\mid$	Ag(s)	E = − 0.70 V
(e)	Al(s)	$\mid$	Al^{3+}(aq)	$\vdots$	Zn^{2+}(aq)	$\mid$	Zn(s)	E = + 0.90 V
(f)	Pb(s)	$\mid$	Pb^{2+}(aq)	$\vdots$	Cd^{2+}(aq)	$\mid$	Cd(s)	E = − 0.27 V

(remember that the sign of the e.m.f. gives the polarity of the right-hand electrode)

20. Construct cell diagrams for the cells in which the following overall reactions occur:

(a) $\quad\quad\quad\quad$ Zn(s) + Cu^{2+}(aq) $\rightarrow$ Zn^{2+}(aq) + Cu(s)

(b) $\quad\quad\quad\quad$ Mg(s) + Pb^{2+}(aq) $\rightarrow$ Mg^{2+}(aq) + Pb(s)

(c) $\quad\quad\quad\quad$ 2Al(s) + $6H^+$(aq) $\rightarrow$ $2Al^{3+}$(aq) + $3H_2$ (g)

(d) $\quad\quad\quad\quad$ Cu(s) + $2Ag^+$(aq) $\rightarrow$ Cu^{2+}(aq) + 2Ag(s)

In each case write the cell diagram so that the right-hand electrode is the more positive.

10
DEDUCTION OF EQUATIONS

1. (a) 0.675 g of aluminium (Al) is found to combine with 2.69 g of chlorine (Cl_2) to give aluminium chloride.

(b) 2.50 g of zinc (Zn) is found to combine with iodine (I_2) to give 12.25 g of zinc iodide.

(c) 4.32 g of silver (Ag) combines with 0.64 g of sulphur (S) to form silver sulphide.

Deduce chemical equations for these reactions.

2. Using the following data deduce equations for the following metal/metal ion displacement reactions.

Wt. of metal added	to excess of	Wt. of metal displaced
(a) 1.50 g zinc	$AgNO_3$ (aq)	4.98 g silver
(b) 1.60 g magnesium	$CuSO_4$ (aq)	4.20 g copper
(c) 3.10 g iron	$Pb(NO_3)_2$ (aq)	12.90 g lead
(d) 0.96 g aluminium	$CuSO_4$ (aq)	3.36 g copper
(e) 1.40 g iron	$CuCl_2$ (aq)	1.57 g copper

3. The following table indicates the volumes of hydrogen evolved when known weights of various metals are treated with excess dilute hydrochloric acid (HCl):

Wt. of metal dissolved	Vol. of hydrogen at s.t.p.
(a) 0.075 g magnesium	70 cm^3
(b) 0.250 g iron	100 cm^3
(c) 0.180 g aluminium	220 cm^3
(d) 0.146 g zinc	50 cm^3

Deduce the equation for each of these reactions.

4. A solution containing 0.526 g of a chloride of chromium was treated with an excess of silver nitrate solution. 1.44 g of silver chloride were precipitated.

What is the formula of the chloride of chromium? Write an equation for the precipitation reaction.

5. When 0.772 g of a chloride of titanium was dissolved in water and added to excess of silver nitrate solution, 2.151 g of silver chloride were precipitated. Deduce the equation for the reaction between the titanium chloride and silver nitrate. (Ti = 48) (L)

6. 1.35 g of caesium chloride were dissolved in water and then excess silver nitrate solution was added. 1.15 g of silver chloride were precipitated. Deduce the formula of caesium chloride and write an equation for its reaction with silver nitrate solution.

7. Heating 2.52 g of orange ammonium dichromate, $(NH_4)_2Cr_2O_7$, produced 1.52 g of a green solid, a condensate of water, and 240 cm^3 of an unreactive gas (measured at $20°C$ and 760 mmHg).

The green solid was found on analysis to have the composition 68.4% chromium and 31.6% oxygen.

What was the formula of the green solid? Write an equation representing the thermal decomposition of ammonium dichromate.

8. Excess sodium hydroxide solution was added to a solution of silver nitrate ($AgNO_3$) and the dark precipitate which was formed was filtered off and dried. 0.232 g of this material yielded on heating 0.216 g of silver and 0.016 g of oxygen. What is the formula of the compound precipitated? Write an equation for the reaction between the two solutions.

If 100 cm^3 of 0.1 M sodium hydroxide solution were added to 25 cm^3 of 0.1 M silver nitrate solution, what volume of 0.1 M hydrochloric acid would be required to neutralise the filtrate left in the experiment? (L)

9. 2.0 g of marble ($CaCO_3$) were added to 100 cm^3 of 0.1 M hydrochloric acid 112 cm^3 of carbon dioxide, measured at s.t.p., were evolved. The unchanged marble was washed and dried and found to weigh 1.5 g.

Deduce an equation for this reaction.

10. Excess sodium hydrogen carbonate solution was added to a solution of lead nitrate when a white precipitate was formed and 110 cm^3 of a colourless gas (measured at s.t.p.) were evolved. This gas turned lime water milky. The precipitate was found to weigh 1.30 g, and on analysis was found to have the percentage composition Pb 77.5%, C 4.5%, O 18.0%.

Deduce an equation for the reaction.

11. Azurite is a blue ore of copper with a formula of the type $xCuCO_3, yCu(OH)_2$.

On heating a sample of azurite 0.246 g of copper (II) oxide were left, 47.5 cm^3 of carbon dioxide were evolved and a condensate of water was obtained. The carbon dioxide was collected at $15°C$ and 760 mmHg pressure.

What was the formula of the azurite? Write an equation for its thermal decomposition.

12. It was found that 10 cm^3 of 2.0 M hydrochloric acid were required to neutralise 20 cm^3 of 0.5 M sodium carbonate solution. In a second experiment 0.318 g of anhydrous sodium carbonate (Na_2CO_3) was treated with excess hydrochloric acid and 71 cm^3 of carbon dioxide were evolved which were collected in a syringe at 760 mmHg and 15°C.

From the results of these two experiments deduce the equation representing the reaction between the sodium carbonate and the hydrochloric acid.

13. When 3.49 g of a chloride of lead were treated with excess of water a dark precipitate was formed. This precipitate was filtered off, washed and dried and was found to weigh 2.40 g. The filtrate was titrated with 1.0 M sodium hydroxide and found to require 40 cm^3 of this alkali for neutralisation. The 2.40 g of dark precipitate were heated in a stream of hydrogen and yielded 2.08 g of lead on complete reduction.

What was the formula of the precipitate? Write an equation for the reaction between the original lead chloride and water.

14. Molar solutions of silver nitrate ($AgNO_3$) and sodium sulphide (Na_2S) were mixed (in test-tubes of equal bore) in the volumes shown in the following table. A black precipitate was formed in each case and all tubes were centrifuged for the same length of time. The heights of the precipitates were measured, and these are given in the table.

Vol. of 1.0 M Na$_2$S (cm^3)	5	5	5	5	5	5	5
Vol. of 1.0 M AgNO$_3$ (cm^3)	2	4	6	8	10	12	14
Height of ppt. formed (mm)	15	30	42	59	72	74	73

Plot a graph of 'height of precipitate' against 'volume of 1.0 M silver nitrate'. Deduce an equation for the reaction.

15. Molar solutions of barium chloride ($BaCl_2$) and sodium carbonate (Na_2CO_3) were mixed (in test-tubes of equal bore) in the volumes shown in the following table. A white precipitate was formed in each case and, after settling, the heights of these precipitates were measured:

Vol. of 1.0 M BaCl$_2$ (cm^3)	10	10	10	10	10	10	10
Vol. of 1.0 M Na$_2$CO$_3$ (cm^3)	2	4	6	8	10	12	14
Height of ppt. formed (mm)	10	21	33	45	58	60	59

Plot a graph of 'height of precipitate' against 'volume of 1.0 M sodium carbonate'. Deduce an equation for the reaction.

16. A 1.0 M solution of lead (II) nitrate and a 0.5 M solution of potassium chromate (K_2CrO_4) were mixed in test-tubes of equal bore in the volumes shown in the following table. The yellow precipitates formed were allowed to settle and their heights measured.

Vol. of 1.0 M $Pb(NO_3)_2$ (cm^3)	5	5	5	5	5	5	5
Vol. of 0.5 M K_2CrO_4 (cm^3)	2	4	6	8	10	12	14
Height of ppt. formed (mm)	4	7	12	17	21	21	20

Plot a graph of 'height of precipitate' against 'volume of 0.5 M potassium chromate'. Deduce an equation for the reaction.

17. When potassium iodide (KI) is added to mercury (II) nitrate ($Hg(NO_3)_2$) a red iodide of mercury is precipitated. However, this red precipitate is soluble in excess potassium iodide solution to form a colourless solution containing a complex ion of the type HgI_x^{n-}

The following table indicates the heights of precipitates obtained when the given volumes of 1.0 M mercury (II) nitrate and 1.0 M potassium iodide were mixed in test-tubes of equal bore.

Vol. of 1.0 M $Hg(NO_3)_2$ (cm^3)	2	2	2	2	2	2	2	2
Vol. of 1.0 M KI (cm^3)	1	2	3	4	5	6	7	8
Height of ppt. formed (mm)	7	15	24	36	23	16	9	0

From the given data deduce the formulae of (i) the red precipitate and (ii) the complex ion, and so write the ionic equations for the reactions involved.

11
DIFFUSION AND EFFUSION OF GASES

1. It was found that 5 cm^3 of a gas diffused in half the time required for the diffusion of 15 cm^3 of oxygen under the same conditions. Calculate the molecular weight of the gas. (L)

2. Calculate the relative rates of diffusion of hydrogen (mol. wt. = 2) and hydrogen iodide (mol. wt. = 128) under similar conditions.

3. Calculate the relative rates of diffusion of hydrogen (H_2) and oxygen (O_2) under similar conditions.

4. A certain gaseous element takes 1.5 times as long to diffuse through a porous plug as does a similar volume of oxygen. Calculate (a) the relative (vapour) density of the gas and (b) its molecular weight. (c) If the gas is diatomic what is its atomic weight?

5. If 100 cm^3 of carbon monoxide diffused through a porous plate in 45 s, how long would it take 150 cm^3 of hydrogen sulphide to pass through the same plate, under similar conditions of temperature and pressure? (L)

6. A hydrocarbon has the percentage composition by weight of 85.7% carbon and 14.3% hydrogen.
53.4 cm^3 of this hydrocarbon diffuse through a porous plug in the same time as 50 cm^3 of oxygen (O_2) under similar conditions.
Calculate (a) the simplest (empirical) formula of the hydrocarbon, (b) its molecular weight. (c) Write its molecular formula.

7. 100 cm^3 of oxygen (O_2) diffuse through a porous partition in 4 min. How long will it take 200 cm^3 of methane (CH_4) to diffuse through the same partition under similar conditions?

8. 90 cm^3 of hydrogen (H_2) diffuse through a porous plug in 6 min. How long would it take 30 cm^3 of hydrogen bromide gas (HBr) to diffuse through the same plug under similar conditions?

9. 104 cm^3 of the noble gas argon effused through a perforated plate in 120 s, while in a similar experiment 87 cm^3 of oxygen (O_2) took 90 s. Calculate the molecular weight of argon. If the atomic weight of argon is 40, write the formula for the argon molecule. The symbol for argon is Ar.

10. Sulphur burns in fluorine to form a fluoride SF$_x$ which is an unreactive gas.

In an effusion experiment it was found that 100 cm^3 of oxygen diffused in 75 s and 50 cm^3 of this sulphur fluoride in 79.5 s, using the same apparatus under similar conditions.

(a) Calculate the molecular weight of the sulphur fluoride. (b) If the atomic weights of sulphur and fluorine are 32 and 19 respectively, what is the formula of the sulphur fluoride? (c) Write the equation for the reaction between sulphur and fluorine.

12
RATES OF REACTION

1. The following graph shows how the rate of decomposition of a sample of hydrogen peroxide varies with time (at $27^{\circ}C$ and 760 mmHg). The initial volume of hydrogen peroxide solution was 50 cm^3 and 2 g of manganese (IV) oxide were used as a catalyst. The reaction was followed by measuring the volume of oxygen evolved after given intervals of time.

$$2H_2O_2 \text{(aq)} \rightarrow 2H_2O \text{(l)} + O_2 \text{(g)}$$

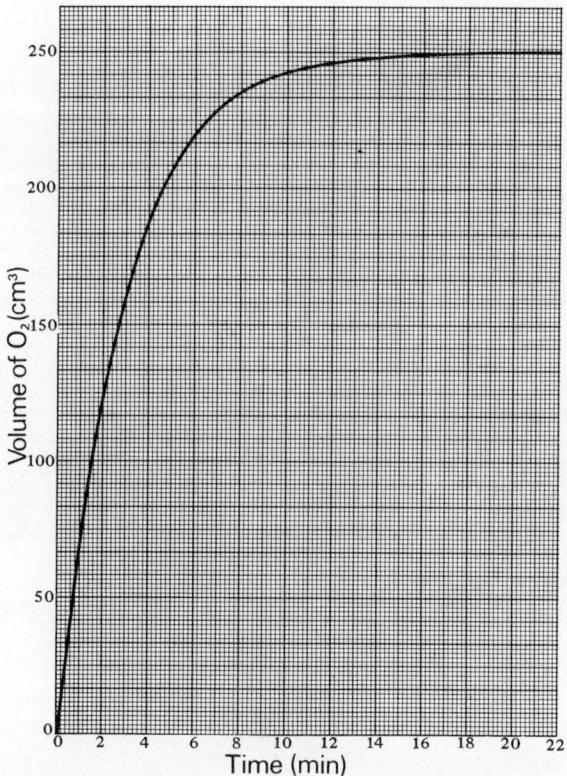

(a) What was the maximum volume of oxygen obtained?

(b) After what time was the reaction complete?

(c) What weight of manganese (IV) oxide would be left at the end of the experiment?

(d) If 1 mole of oxygen (O_2) occupies 25 dm^3 at 27°C and 760 mmHg, how many moles of oxygen were formed when the reaction was complete?

(e) How many moles of hydrogen peroxide (H_2O_2) were present in the original 50 cm^3?

(f) What was the concentration of the original hydrogen peroxide solution in

(i) mol dm^{-3} (ii) g dm^{-3}.

(g) What fraction of the hydrogen peroxide had decomposed after 4.0 min?

(h) Copy the curve on to a sheet of graph paper; then draw the (approximate) curves that you would expect the experimenter to have obtained if (i) the hydrogen peroxide solution had been half the concentration initially, (ii) the initial hydrogen peroxide concentration had been the same but the temperature about 10°C higher.

2. In a series of experiments in which magnesium ribbon of uniform width reacted with excess dilute hydrochloric acid, the rates of evolution of hydrogen were found to be as follows:

Length of ribbon (cm)	1.0	2.0	3.0	4.0	5.0	6.0	7.0
Rate of evolution of hydrogen (cm^3 min^{-1})	1.1	1.8	2.7	3.6	4.6	5.4	6.4

Draw a graph and state what conclusion you reach. Calculate the rate of evolution of hydrogen from a piece of magnesium ribbon 12 cm long under the same conditions.

3. 25 cm^3 of 1.0 M hydrochloric acid were added to an excess of marble chips in a flask. The neck of the flask was plugged with cotton wool and the flask and contents were placed on a balance. The weight was noted initially and after intervals of time until the reaction appeared to have ceased. Results from the experiment are given in the following table:

Weight (g)	159.66	159.50	159.38	159.30	159.24	159.17
Time (min)	0	0.5	1.0	1.5	2.0	3.0

159.14	159.12	159.11	159.10	159.10	159.10
4.0	5.0	6.0	7.0	8.0	9.0

The equation for the reaction is

$$CaCO_3 (s) + 2HCl(aq) \rightarrow CaCl_2 (aq) + CO_2 (g) + H_2O(l)$$

Plot a graph of weight against time and use this graph to answer the following:

(a) After what time was the reaction half-way to completion?

(b) What weight of carbon dioxide had been evolved after 8.0 min? What volume would this weight of gas occupy at s.t.p?

(c) On the same piece of graph paper as your plotted curve draw two other curves illustrating the results that you would expect the experimenter to have obtained had the initial acid been (i) 25 cm^3 of 0.5 M HCl, (ii) 25 cm^3 of 2.0 M HCl.

4. 25 cm^3 of 0.1 M sulphuric acid were added to an excess of freshly cleaned, granulated zinc in a reaction vessel. The vessel was connected to a gas burette so that the hydrogen evolved by the reaction

$$Zn(s) + H_2SO_4(aq) \rightarrow ZnSO_4(aq) + H_2(g)$$

could be collected and measured after intervals of time. The following table gives the results obtained.

Vol. of hydrogen (cm^3)	0	30	45	52.5	56.3	58.2	60.0
Time (min)	0	1	2	3	4	5	6

60.4	60.6	60.7	60.8	60.8	60.8
7	8	9	10	11	12

The room temperature and pressure at the time of the experiment were 21°C and 755 mmHg respectively.
(a) Plot the curve and explain its shape (i.e. why does it commence steeply, and slowly level off until it is horizontal?).
(b) Show that the maximum volume of hydrogen obtained is in reasonable agreement with the volume you would obtain by calculation using the above equation.
(c) By drawing a tangent to the curve and measuring its slope obtain a value for the rate of reaction after 2 min have elapsed. Express the rate as a rate of evolution of hydrogen (cm^3 min^{-1}) under room conditions.

5. The graph on page 48 summarises experimental results obtained from the radioactive disintegration of an isotope of thallium, $^{206}_{81}$Tl. The y-axis represents the number of grammes of this isotope present in the sample.

(a) During which of the following time intervals is the average rate of disintegration greatest?
   (i) 0 – 2 min, (ii) 8 – 10 min, (iii) 18 – 20 min.
(b) After what interval of time had half of the original number of atoms disintegrated? (i.e. what is the 'half-life' of this radioisotope?)
(c) How long did it take for one-quarter of the isotope to disintegrate?
(d) What fraction of the original quantity of isotope was left after 6 minutes?
(e) How long did it take the isotope to decrease from
   (i) $\frac{1}{2}$ to $\frac{1}{4}$ of its original amount? (ii) $\frac{1}{4}$ to $\frac{1}{8}$? (iii) $\frac{1}{8}$ to $\frac{1}{16}$?
(f) If twice as much of the isotope had been present initially, how long would it have taken for half of this quantity to disintegrate?

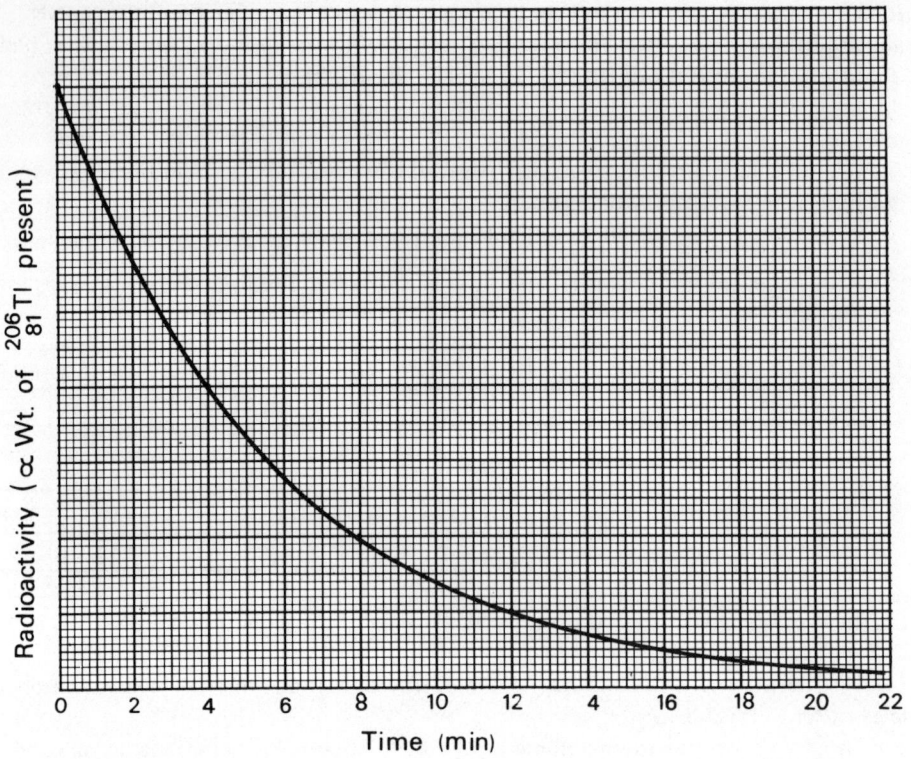

Radioactivity (∝ Wt. of $^{206}_{81}$Tl present)

Time (min)

6. 40 cm^3 of 1.0 M hydrochloric acid were added to 20 g (a large excess) of marble chips in a conical flask. The flask and contents were immediately weighed and a stop-clock started. The weight of flask and contents was noted as the reaction proceeded and the following table indicates the loss in weight at various times.

Time (min)	0	0.5	1.0	1.5	2.0	3.0	4.0
Loss in weight (g)	0	0.19	0.35	0.47	0.56	0.69	0.76

5.0	6.0	7.0	8.0	9.0	10.0
0.80	0.83	0.84	0.86	0.86	10.86

Plot a graph of 'loss in weight' against 'time', and use your graph to answer the following questions:
(a) How long did it take for half the acid to be used up?
(b) After what time was the reaction virtually complete?
(c) How long did it take for the second half of the acid to be used up?
(d) Why did the second half of the reaction take longer than the first half?
(e) What weight of carbon dioxide had been evolved when the reaction was complete?

(f) How many moles of carbon dioxide were evolved?
(g) How many moles of calcium carbonate ($CaCO_3$) (marble chips) must have reacted in order to produce this number of moles of carbon dioxide?
(h) How many moles of hydrochloric acid (HCl) were there at the beginning of the reaction?
(i) Write the equation for the reaction between marble and hydrochloric acid.

7. (This question also uses the data of question 6, which should be attempted before this one.)
The experiment in question 6 was repeated, except that the concentration of the hydrochloric acid was increased to 2.0 M. The results were as follows

Time (min)	0	0.5	1.0	2.0	3.0	4.0	5.0
Loss in weight (g)	0	0.47	0.83	1.23	1.45	1.57	1.63

6.0	7.0	8.0	9.0	10.0	11.0
1.66	1.68	1.70	1.72	1.72	1.72

Plot a graph of these results similar to that in question 6. Use the graph to answer the following questions:
(a) How many moles of carbon dioxide were evolved during this reaction?
(b) Show that this quantity of carbon dioxide (allowing for experimental error) is the quantity you would expect from the equation written in question 6.
(c) Measure the slope of this graph and also that of the curve obtained in question 6, both at time = 1 min. How do these slopes relate to the concentrations of the acid used in the two experiments? Of what may the slopes of the graphs be taken as a measure?

8. Square-shaped pieces of aluminium foil, of different sizes, were treated separately with hydrochloric acid. The reaction which occurred may be represented by the equation
$$2Al(s) + 6H^+(aq) \rightarrow 2Al^{3+}(aq) + 3H_2(g)$$

The rate at which hydrogen was evolved was measured in each case, and the following table summarises the results.

Length of side of square of foil (cm)	1.0	1.3	1.8	2.1	2.5	2.9
Rate of formation of hydrogen (cm^3 min^{-1})	2.0	3.4	6.5	8.8	12.5	16.8

Show graphically that the rate of production of hydrogen is directly proportional to the surface area of the foil.
What would be the rate of production of hydrogen using a piece of foil (a) of dimensions 2.0 cm $\times$ 2.5 cm, (b) of surface area 16.0 cm^2, and acid of the same concentration?

9. A piece of tinfoil suspended from the beam of a balance was completely immersed in a 1.0 M solution of iodine in toluene. The tin reacted slowly with the iodine

$$Sn(s) + 2I_2 \text{(toluene)} \rightarrow SnI_4 \text{(toluene)}$$

and the change in weight was noted after intervals of time up to 15 min. The reaction was slow and only small proportions of the reactants had been consumed at the end of this period.

The experiment was repeated four times with similar pieces of foil, but the concentrations of the iodine solutions were progressively reduced. The following table gives the results of the experiments. It may be assumed that the temperature was the same for all the experiments.

Time (min)	Weight of piece of tin foil (g)				
	Expt. 1 (1.0 M I_2)	Expt. 2 (0.8 M I_2)	Expt. 3 (0.6 M I_2)	Expt. 4 (0.4 M I_2)	Expt. 5 (0.2 M I_2)
0	3.200	3.200	3.200	3.200	3.200
1	3.187	3.189	3.192	3.194	3.195
2	3.173	3.178	3.184	3.189	3.194
4	3.147	3.157	3.168	3.178	3.189
7	3.106	3.125	3.144	3.163	3.181
10.5	3.060	3.088	3.116	3.144	3.172
15	3.001	3.040	3.079	3.120	3.159

(a) Plot curves of 'weight of tin foil' against 'time' for each experiment — preferably on the same piece of graph paper.

(b) From your graphs obtain values for the rate of each reaction in terms of 'change of weight per minute'.

(c) Plot another graph of 'rate of reaction' against 'iodine concentration'. What conclusion can you draw from this graph?

10. In a reaction between a solid D and a solution of E a gas is evolved. The reaction was performed three times with the following quantities of reactants (D being in excess each time):

(i)   5g of D and 25 cm^3 of a 0.2 M solution of E

(ii) 10 g of D and 25 cm^3 of a 0.2 M solution of E

(iii) 5 g of D and 25 cm^3 of a 0.1 M solution of E

In each case the volume of gas liberated at various times after the start was noted. The graph on page 51 has three curves. A, B and C which summarise the three sets of results. Which of the above experiments is responsible for each curve? (Question continued on page 52.)

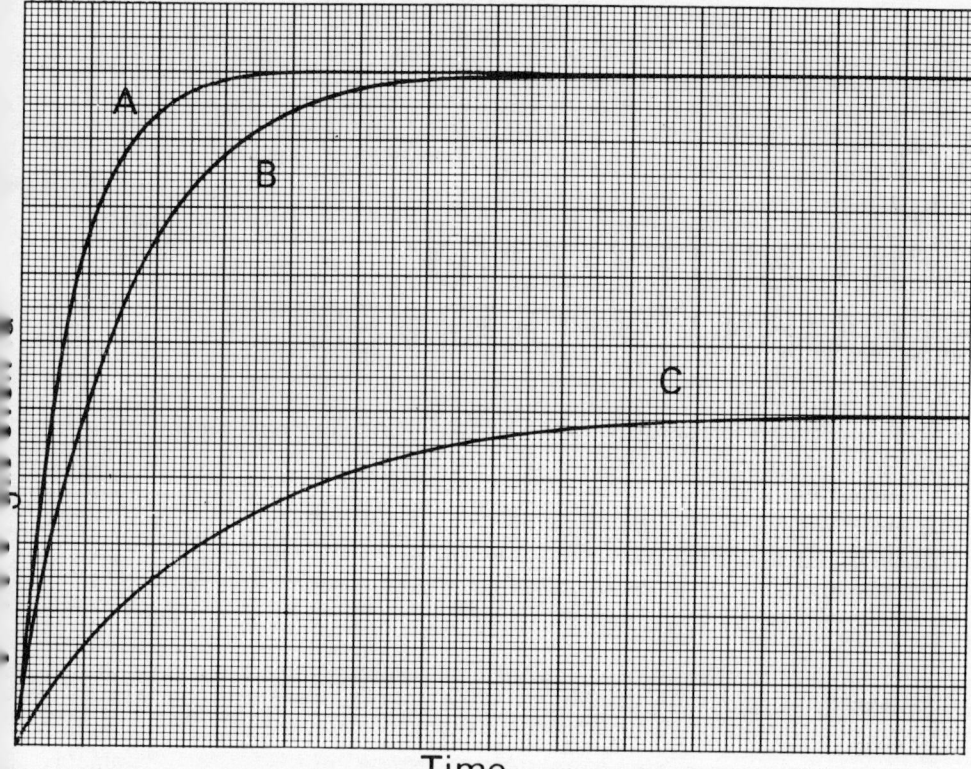

Time

On a sheet of graph paper, but using the same axes as shown, draw the curves that you would expect to obtain if the experiment were performed with the following quantities of reactants:

(iv) 10 g of D and 25 cm³ of a 0.1 M solution of E
(v)   10 g of D and 50 cm³ of a 0.1 M solution of E.

11. The reaction between sodium thiosulphate ($Na_2S_2O_3$) and hydrochloric acid is

$$S_2O_3{}^{2-}(aq) + 2H^+(aq) \rightarrow S(s) + SO_2(aq) + H_2O(l)$$

and the time taken for a precipitate of sulphur to appear may be taken as an inverse measure of the rate of the reaction.

In each of six experiments a volume (V cm³) of a given solution of thiosulphate was diluted to 50 cm³ with water and then treated with 5 cm³ of 2.0 M hydrochloric acid. V (and therefore the thiosulphate concentration) was varied. The time for a precipitate to appear (t seconds) was noted in each case. The results were

Vol. of thiosulphate (V cm³)	50	40	30	20	15	10
Vol. of water added (cm³)	0	10	20	30	35	40
Time (t seconds)	33	42	54	76	100	178
$1/t \times 10^3$ s^{-1}	30	25	18.5	13	10	5.6

(a) Plot a graph of V against t. (V may be taken as a measure of thiosulphate solution.)
(b) Plot a graph of V against $1/t$ ($1/t$ being a measure of reaction rate).

From your graphs, what may you deduce about the relationship between reaction rate and concentration of thiosulphate?

13
SOLUBILITY AND SOLUBILITY CURVES

1. In an experiment to determine the solubility of ammonium bromide in water the following results were obtained:

Weight of dish	= 25.25 g
Weight of dish + saturated solution at 20°C	= 42.72 g
Weight of dish + solid ammonium bromide after evaporation to dryness	= 32.72 g

Calculate the solubility of ammonium bromide as grammes per 100 g water at 20°C.

2. The results of an experiment to determine the solubility of potassium chlorate in water at 30°C were as follows:

Weight of dish	= 15.86 g
Weight of dish + saturated solution at 30°C	= 26.86 g
Weight of dish + solid potassium chlorate after evaporation to dryness	· = 16.86 g

(a) Calculate the solubility of potassium chlorate in grammes per 100 g water at 30°C.

(b) What would be the weight of a saturated solution containing 50 g of water, also at this temperature?

3. The solubility of potassium dichromate at 20°C is 12 g per 100 g of water, and at 80°C it is 50 g per 100 g of water.

(a) What weight of potassium dichromate is present in 70 g of saturated solution at 80°C?

(b) What weight of potassium dichromate would crystallise if the solution in (a) were cooled to 20°C?

(c) What weight of saturated solution remains at 20°C?

4. 100 g of water at 15°C dissolve at saturation 37 g of sodium chloride and 25 g of potassium nitrate, but at 70°C the corresponding weights are 38 g and 140 g per 100 g of water respectively.

100 g of a mixture of the above two salts, in equal proportions by weight, are shaken with 100 g of water at 70°C till equilibrium is reached. The solution is filtered hot. The

hot filtrate is slowly cooled to 15°C and again filtered. The final filtrate is evaporated to dryness.

What will be the weights and compositions of the three residues, assuming that no solution is left in either of the filter papers? (C)

5. The following figures were found for the solubility of a salt at various temperatures, the solubility being expressed as grammes per 100 g of water.

Temperature (°C)	10	20	30	40	50	60
Solubility	17.7	24.1	31.6	39.2	46.3	57.5

Plot the solubility curve and use this to find the solubility of this salt at (a) 35°C and (b) 45°C.

6. The solubility of ammonium bromide (expressed as grammes of anhydrous salt, NH_4Br, per 100 g of water) at various temperatures is given in the following table.

Temperature (°C)	20	30	40	50	60	70	80
Solubility	75	83	91	99	108	117	126

Plot the solubility curve between 20°C and 80°C.
Use your graph to answer the following questions:
(a) What is the solubility of ammonium bromide at (i) 45°C, (ii) 66°C?
(b) A saturated solution of ammonium bromide in 100 g of water at 70°C was allowed to cool to 37°C. What weight of ammonium bromide crystals was formed?
(c) At what temperature is the solubility of ammonium bromide 105 g per 100 g of water?
(d) A solution contains 90 g of ammonium bromide per 100 g water at 55°C.
  (i) What further weight of solid ammonium bromide has to be added to this solution to make it saturated?
  (ii) If the above solution were cooled, at what temperature would crystals first appear (assuming no supercooling)?

7. The solubility of sodium nitrate in water (expressed as grammes of anhydrous salt per 100 g of water) at various temperatures is given in the following table.

Temperature (°C)	10	20	30	40	50	60	70
Solubility	80	88	96	105	114	124	135

Plot the solubility curve between 10°C and 70°C.
Use your graph to answer the following questions:

(a) What is the solubility of sodium nitrate at (i) 25°C, (ii) 55°C?

(b) If a saturated solution of sodium nitrate in 100 g of water at 55°C, were cooled to 25°C, what weight of sodium nitrate would crystallise?

(c) At what temperature is the solubility of sodium nitrate 100 g per 100 g water?

(d) If 60 g of sodium nitrate were added to 50 g of water and this mixture was maintained at 45°C till equilibrium was established, what weight of salt would remain undissolved?

(e) State whether the following mixtures of sodium nitrate and water give unsaturated or saturated solutions:

    (i)   85 g sodium nitrate, 100 g water at 25°C

    (ii)   98 g sodium nitrate, 100 g water at 50°C

    (iii) 125 g sodium nitrate, 100 g water at 55°C

    (iv) 125 g sodium nitrate, 100 g water at 65°C

    (v)   50 g sodium nitrate, 50 g water at 30°C.

8. The solubilities of two salts D and E are given in the following table. In each case the solubility is expressed as grammes per 100 g water.

Temperature (°C)	10	20	30	40	50	60	70	80
Solubility of D	17	21	24	29	34	40	47	56
Solubility of E	35.8	36	36.2	36.5	36.8	37.3	37.6	38.0

Using these data, plot solubility curves for D and E on the same sheet of graph paper. Use the graphs you have drawn to answer the following:

(a) At what temperature are the solubilities of the two salts equal?

(b) Estimate the solubility of D at 0°C.

(c) A saturated solution of E in 50 g of water at 25°C was evaporated to dryness. What was the weight of the residue?

(d) Two separate 100 g portions of water are saturated at 75°C, one with D and the other with E. What is the difference in weight between these two solutions?

(e) If the two solutions in (d) are cooled to 25°C what weight of crystals will separate from each?

9. The following table shows the solubilities of two salts, A and B, at various temperatures, expressed as grammes of salt per 100 g of water. Plot the two solubility curves on one piece of graph paper.

Temperature (°C)	0	10	20	30	40	50	60	70	80
Solubility of A	28	31	34	37	40	43	45	48	51
Solubility of B	13	21	32	46	64	85	110	138	169

From your graph answer the following questions:

(a) Above what temperature is B the more soluble salt?

(b) What happens when a mixture of 100 g of B with 100 g of water is heated to 80°C?

(c) What happens when the mixture in (b) is cooled from 80°C to 20°C?

(d) A mixture of 40 g of A and 60 g of B is added to 100 g of water and heated to 70°C until there is no further change, and the solution is then cooled to 10°C. Describe what will happen and calculate the percentage of A and of B in the crystals that are formed.

(L)

10. The solubility of gaseous carbon dioxide, expressed as grammes per 100 g of water, is given at various temperatures and at two different pressures, 1 atm and 0.5 atm.

Temperature °C	0	10	20	30	40	50	60
Solubility at 1 atm	0.34	0.24	0.17	0.13	0.10	0.075	0.055
Solubility at 0.5 atm	0.17	0.12	0.09	0.07	0.05	0.04	0.03

Plot the solubility curves for carbon dioxide in water at (i) 1 atm pressure, (ii) 0.5 atm pressure, on the same sheet of graph paper.

Use your graph to answer the following questions:

(a) How does the solubility of carbon dioxide change as temperature increases, the pressure remaining constant?

(b) What is the solubility of carbon dioxide in water at (i) 5°C and 1 atm, (ii) 5°C and 0.5 atm?

(c) 200 g of water were saturated with carbon dioxide at 35°C and 1 atm. (i) What weight of gas dissolved? (ii) What volume of gas, expressed at s.t.p., dissolved?

(d) How does the solubility of carbon dioxide change as pressure increases, the temperature remaining constant?

ANSWERS

1. THE MOLE CONCEPT (p. 1.)

1. (a) 32   (b) 44   (c) 96   (d) 101   (e) 46   (f) 278
2. (a) 4.0   (b) 0.5   (c) 1.0   (d) 0.25   (e) 0.845   (f) 9
3. (a) 4   (b) 0.7   (c) 0.4   (d) 0.1   **4.** (a) 0.4   (b) 7.5   (c) 1.0   (d) 0.2
5. (a) 6 g   (b) 3.65 g   (c) 1.76 g   (d) 0.017 g   (e) 156.6 g   (f) 68 g
(g) 20 g   (h) 17 g   **6.** (a) 108 g   (b) 24 g   (c) 3.9 g   (d) 2.8 g   (e) 54 g
7. (a) 0.5   (b) 2.0   (c) 0.1   (d) 0.05   (e) 0.05
8. (a) 0.75 mole of iron react with 0.5 mole of oxygen to produce an oxide of iron.
(b) 0.25   (c) 3:2:1   **9.** (a) (i) 0.5 (ii) 0.75   (b) (i) 3.1 (ii) 6.2 (iii) 12.4
10. 2   **11.** (a) PCl_3 (b) $CaCO_3$ (c) PbO_2 (d) CH (e) NH_4Cl
(f) $Na_2S_2O_3$ (g) C_2H_5 **12.** $Na_2SO_4, 10H_2O$ **13.** $ZnSO_4, 7H_2O$
14. 52.9%   **15.** 11.25%
16. (a) 86.6%   (b) 21.2%   (c) 25.45%   (d) 60.7%   (e) 70.0%
17. (a) 62.9%   (b) 36.1%   (c) 45.3%   (d) 51.2%   (e) 45.6%

2. THE GAS LAWS (p. 4.)

1. (a) 290 K   (b) 298 K   (c) 260 K   (d) 73 K   (e) 573 K   (f) 244 K
2. (a) 27°C   (b) −61°C   (c) −2°C   (d) −223°C   (e) 381°C   (f) −273°C
3. (a) 58.8 cm³   (b) 24.44 dm³   (c) 27.08 cm³   (d) 3.69 dm³
4. (a) 33.2 cm³   (b) 118.7 cm³   (c) 3532 dm³   (d) 25.94 cm³
5. (a) 317 cm³   (b) 241 cm³   (c) 443 cm³   (d) 0.851 dm³

3. MOLAR VOLUMES AND RELATIVE (VAPOUR) DENSITY (p. 5.)

1. (a) to (e) 22.4 dm³
2. (a) 22.4 dm³   (b) 11.2 dm³   (c) 44.8 dm³   (d) 2.24 dm³   (e) 0.56 dm³
3. (a) 30   (b) 34   (c) 16.1   (d) 64.1   (e) 46.1   **4.** (a) 34   (b) 5 g
5. 44   **6.** 34
7. (a) 10 cm³   (b) 39.0 cm³   (c) 18.3 cm³   (d) 13.8 cm³   (e) 6.63 cm³
8. (a) 12.8 cm³   (b) 23.7 cm³   (c) 44.9 cm³
9. (a) 96.9 cm³   (b) 58.2 cm³   (c) 18.0 cm³   (d) 62.3 cm³
10. (a) LiCl 20.5 cm³; NaCl 27.1 cm³; KCl 37.6 cm³; RbCl 43.2 cm³; CsCl 42.1 cm³
11. (a) NaF 16.4 cm³; NaCl 27.1 cm³; NaBr 32.2 cm³; NaI 40.9 cm³
12. (a) 115 cm³   (b) 130 cm³   (c) 146 cm³   (d) 162 cm³   (e) 178 cm³
13. (a) 32   (b) 14   (c) 22   (d) 8.5   (e) 17   (f) 35.5   (g) 20   (h) 64   (i) 14.4
14. (a) $C_2H_4Br_2$ (b) C_6H_6 (c) CH_2O (d) $C_2H_4O_2$ (e) S_2Cl_2
15. (a) 29   (b) 58   (c) C_4H_{10}

4. REACTING WEIGHTS (not including gas volumes) (p. 9.)

1. (a) 1.45 g   (b) 4.02 g   (c) 2.23 g   (d) 1.59 g   (e) 1.60 g
2. 0.515 g   **3.** 1.39 g   **4.** 32.5 g dm⁻³   **5.** 1.68 tons   **6.** 2.76 tons
7. 9.3 g   **8.** 7.6 g, 0.05 mole   **9.** (i) 85 tons   (ii) 315 tons   **10.** 10.95 g
11. (i) 7.18 g   (ii) 3.41 g   **12.** 38.5% Fe; 61.5% Fe_2O_3

13. 54.7% $ZnCO_3$; 45.3% ZnO **14.** 9.1% $NaNO_3$; 90.9% Na_2SO_4 **15.** 33.3%
16. 50%, 0.39 g   **17.** 10.54 g   **18.** Na_2CO_3, H_2O **19.** 1 : 2
20. $Cu(OH)_2$, $2CuCO_3$ **21.** (a) Na_2CO_3, $NaHCO_3$, $2H_2O$ (b) 0.292 g

5. REACTING WEIGHTS (including gas volumes) (p. **12.**)

1. (a) 112 cm³   (b) 1.12 dm³   (c) 560 cm³   (d) 1.79 dm³
2. (a) 1.18 dm³   (b) 122 cm³   (c) 3.56 dm³   (d) 4.77 dm³   (e) 6.46 dm³
3. (a) 2.24 dm³   (b) 11.7 g   **4.** 14.11 g, 12.68 g   **5.** 11.2 dm³
6. 11.2 dm³   **7.** (i) 373 dm³   (ii) 2566 dm³   **8.** 1 : 2.47 (0.405 : 1)
9. (a) 1.8 g   (b) 1.12 dm³   **10.** x = 10
11. (a) 224 cm³   (b) 238 cm³   (c) 0.03 mole   **12.** 431 cm³   **13.** 8 dm³
14. (a) Li_3N (b) 1   (c) $Li_3N + 3H_2O \rightarrow 3LiOH + NH_3$ (d) Ca_3N_2
15. (i) 402 cm³   (ii) 804 cm³   **16.** $2PbCO_3$, $Pb(OH)_2$

6. REACTING VOLUMES OF GASES (p. **15.**)

1. (a) 1.0 dm³ H_2, 2.0 dm³ HCl   (b) 20 cm³ NO, 20 cm³ NO_2
(c) 100 cm³ H_2S, 150 cm³ O_2, 100 cm³ $H_2O(g)$ (d) 25 cm³ CO_2
(e) 32 cm³ C_4H_{10}, 128 cm³ CO_2, 160 cm³ $H_2O(g)$ (f) 7.5 cm³ H_2S
2. (a) $2H_2(g) + O_2(g) \rightarrow 2H_2O(g)$; and $2CO(g) + O_2(g) \rightarrow 2CO_2(g)$ (b) 23.5 cm³
(c) 117.5 cm³
3. (i) 50 cm³ CO_2, 30 cm³ O_2 (ii) 50 cm³ CO_2, 30 cm³ O_2, 60 cm³ $H_2O(g)$
4. 3287.5 dm³   **5.** All O_2, 20 cm³   **6.** 100 cm³   **7.** 50%
8. 10 cm³ O_2, 12 cm³ CO_2, 20 cm³ N_2 **9.** 8 cm³ NH_3, 12 cm³ N_2
10. 10% N_2, 90% NO   **11.** 20.8%   **12.** (a) 50% CO, 50% H_2 (b) 50 cm³
13. 28 cm³ CO, 12 cm³ CO_2 **14.** 16 cm³ CO, 14 cm³ CO_2, 20 cm³ N_2
15. 3% CO_2, 30% O_2, 67% N_2 **16.** x = 4   **17.** x = 2
18. (a) 2   (b) 64   (c) XO_2
19. (a) C_2N_2 (b) $C_2N_2(g) + 2O_2(g) \rightarrow 2CO_2(g) + N_2(g)$ **20.** SbH_3
21. C_4H_{10} **22.** N_2O
23. (i) 0.125   (ii) 56, 84   (iii) C_4H_8 (iv) 1
(v) $C_4H_8(g) + 6O_2(g) \rightarrow 4CO_2(g) + 4H_2O(l)$ (vi) 60 cm³ O_2, 40 cm³ CO_2

7. STANDARD SOLUTIONS (p. **19.**)

1. (a) 5.85 g   (b) 170 g   (c) 12.48 g   (d) 245.1 g   (e) 86.7 g   (f) 111.0 g
2. (a) (i) 0.1 M, (ii) 0.1 M   (b) (i) 0.6 M, (ii) 0.9 M   (c) (i) 2.4 M, (ii) 1.2 M
(d) (i) 2.4 M, (ii) 4.8 M
3. (a) 106 g   (b) 12.48 g   (c) 0.429 g   (d) 6.8 g   (e) 49.95 g   (f) 5.75 g
4. (a) 0.1 M   (b) 0.5 M   (c) 0.048 M   (d) 1.59 M   (e) 0.1 M
5. (a) 67.6 g   (b) 49 g   (c) 1.03 g   (d) 268.5 g   (e) 17.9 g
6. 12.5 cm³   **7.** 17.5 M   **8.** (a) 0.0928 M   (b) 3.712 g
9. (a) 0.233 M   (b) 13.05 g   **10.** 0.106 M, 7.87 g
11. (a) 0.848 M   (b) 14.42 g   **12.** 0.1 M, 0.1 g-ion   **13.** x = 10
14. 15.9 g dm⁻³   **15.** (a) 320 cm³   (b) 998 cm³
16. (a) 28 cm³   (b) 1.12 M   (c) 44.8 g   (d) 0.5 M
17. molarity = 1.9 M, 107 g dm⁻³ KOH   **18.** 0.08 M
19. $CeCl_4$ $Ce(s) + 2Cl_2(g) \rightarrow CeCl_4(s)$
20. Number moles zinc required = 1 $CuSO_4(aq) + Zn(s) \rightarrow Cu(s) + ZnSO_4(aq)$

8. THERMOCHEMISTRY (p. 25.)

1. (a) $3930\,kJ$ ($940\,kcal$)　(b) $98.25\,kJ$ ($23.5\,kcal$)　(c) $393\,kJ$ ($94\,kcal$)
(d) $589.5\,kJ$ ($141\,kcal$)　(e) $6\,g$　(f) $24\,g$　(g) $16.6\,g$
2. (a) endothermic　(b) $+6.3\,kJ$ ($+1.5\,kcal$)　(c) $+12.6\,kJ$ ($+3.0\,kcal$)
(d) $+12.6\,kJ$ ($+3.0\,kcal$)
3. $\Delta H = -155\,kJ$ ($-37\,kcal$) $g-eq^{-1}$
4. (a) exothermic　(b) (i) 1　(ii) 2　(c) $-300\,kJ$ ($-72\,kcal$)
(d) (i) $49.1\,g$　(ii) $1.5\,g$　(e) 1 mole
5. $-393.2\,kJ$ ($-94\,kcal$)
$$C(s)+O_2(g) \rightarrow CO_2(g); \Delta H = -393.2\,kJ \ (-94\,kcal) \ g-eq^{-1}$$
6. $-1336\,kJ$ ($-320\,kcal$)
$$CH_4(g)+2O_2(g) \rightarrow CO_2(g)+2H_2O; \Delta H = -1336\,kJ \ (-320\,kcal) \ g-eq^{-1}$$
7. (a) $\Delta H = -1814\,kJ$ ($-422\,kcal$) $g-eq^{-1}$
$$\Delta H = -3628\,kJ \ (-844\,kcal) \ g-eq^{-1}$$
(b) $\Delta H = -3000\,kJ$ ($-720\,kcal$) $g-eq^{-1}$　　$1500\,kJ$ ($360\,kcal$)
(c) $\Delta H = -83.6\,kJ$ ($-20\,kcal$) $g-eq^{-1}$　(d) $\Delta H = +115\,kJ$ ($+27.5\,kcal$) $g-eq^{-1}$
8. $+6.40\,kJ$ ($+1.53\,kcal$)　　**10.** $-109\,kJ$ ($-26\,kcal$)
11. (a) $+17\,kJ$ ($+4\,kcal$) $g-eq^{-1}$　(b) endothermic　(c) yellow　(d) red
12. $\Delta H = -5152\,kJ$ ($-1232\,kcal$) $g-eq^{-1}$　　$a = 12, b = 10, c = 4$
13. $\Delta H = -2015\,kJ$ ($-482\,kcal$) mol^{-1}　　**14.** $-2680\,kJ$ ($-641\,kcal$) mol^{-1}
15. (a) $26.3\,kJ$ ($6.3\,kcal$)　(b) $-365.9\,kJ$ ($-87.6\,kcal$)
(c) $S(s)+O_2(g) \rightarrow SO_2(g); \Delta H = -365.9\,kJ$ ($-87.6\,kcal$) $g-eq^{-1}$
16. (a) $18.75\,kJ$ ($4.48\,kcal$)　(b) $-394.7\,kJ$ ($-94.4\,kcal$)
(c) $C(s)+O_2(g) \rightarrow CO_2(g); \Delta H = -394.7\,kJ$ ($-94.4\,kcal$) $g-eq^{-1}$
17. $\Delta H = +25.9\,kJ$ ($+6.2\,kcal$) $g-eq^{-1}$
18. $1.0\,M$ (in equation); $\Delta H = -66.9\,kJ$ ($-16\,kcal$) $g-eq^{-1}$
19. $\Delta H = -1137\,kJ$ ($-272\,kcal$) $g-eq^{-1}$　　**20.** $\Delta H = -21.7\,kJ$ ($-5.2\,kcal$) $g-eq^{-1}$
21. $\Delta H = -19.2\,kJ$ ($-4.6\,kcal$) $g-eq^{-1}$; $2.3°C$
22. $\Delta H = -83.6\,kJ$ ($-20\,kcal$) $g-eq^{-1}$　(a) $3.3°C$　(b) $2.7°C$　(c) $5.0°C$
(d) $3.3°C$
23. (a) $H^+(aq)+OH^-(aq) \rightarrow H_2O(l); \Delta H = -56.8\,kJ$ ($-13.6\,kcal$) $g-eq^{-1}$
24. (b) $-5270\,kJ$ ($-1150\,kcal$) mol^{-1}　(c) 10 carbon atoms, $C_{10}H_{22}$
25. (a) $-41.2\,kcal$　(c) $\Delta H = -38.2\,kcal \ g-eq^{-1}$
26. (a) $3.52\,kJ$ ($0.84\,kcal$)　(b) $1.17\,kJ$ ($0.28\,kcal$) min^{-1}
27. (a) $4.18\,kJ$ ($1\,kcal$) min^{-1}　(b) $41.8\,kJ$ ($10\,kcal$)　(c) $37.6\,kJ$ ($9\,kcal$)
28. (a) $4.68\,kJ$ ($1.13\,kcal$) min^{-1}　(b) $46.8\,J$ ($11.3\,cal$)　(c) $41.97\,kJ$ ($10.14\,kcal$)
29. (a) $27.6\,kJ$ ($6.6\,kcal$) mol^{-1}　(b) $29.3\,kJ$ ($7.0\,kcal$) mol^{-1}
(c) $30.5\,kJ$ ($7.3\,kcal$) mol^{-1}　(d) $33.5\,kJ$ ($8.0\,kcal$) mol^{-1}
(e) $37.7\,kJ$ ($9.0\,kcal$) mol^{-1}　(f) $41.0\,kJ$ ($9.8\,kcal$) mol^{-1}

9. ELECTROCHEMISTRY (p. 34.)

1. (a) 1　(b) 2　(c) 4　(d) 2　(e) 1　(f) 1
2. (a) 0.05　(b) 0.033　(c) 0.10　(d) 0.05　(e) 0.05　(f) 0.10
3. (a) $+2$　(b) $+1$　(c) -1　(d) $+2$　　**4.** M_2O_3, MCl_3, $M_2(SO_4)_3$
5. (a) 2　(b) 0.0125　(c) $20.7\,g$　(d) 0.00623　　**6.** $3.40\,g$
7. $0.567\,g$ Cu, $1.93\,g$ Ag　　**8.** $168\,cm^3$　　**9.** 1 h 20 min　　**10.** $1.05\,A$
11. $2.75\,A$　　**12.** (a) 0.0187　(b) 0.0047　(c) $105\,cm^3$　　**13.** 2 F; 1 F

14. (a) 1:3   (b) 0.0022 F   (c) (i) 25 cm³ (ii) 50 cm³

15. (a) 0.2 amp-hours (720 coulombs)   (b) 65.8 g   (c) 3   (d) 3+

16. (a) 0.005   (b) 0.0025   (c) 2   (d) Sn^{4+} (e) $2Fe^{3+} + Sn^{2+} \rightarrow 2Fe^{2+} + Sn^{4+}$
(f) 0.05

17. (a) −0.76 V, Pt   (b) +1.10 V, Cu   (c) −0.93 V, Ag   (d) +0.34 V, Cu
(e) −0.46 V, Ag   (f) +1.56 V, Ag

18. (a) (i) $2H^+(aq) + 2e \rightarrow H_2(g)$ (ii) $Zn(s) \rightarrow Zn^{2+}(aq) + 2e$
(iii) $Zn(s) + 2H^+(aq) \rightarrow Zn^{2+}(aq) + H_2(g)$ (b) (i) $Cu^{2+}(aq) + 2e \rightarrow Cu(s)$
(ii) $Zn(s) \rightarrow Zn^{2+}(aq) + 2e$ (iii) $Cu^{2+}(aq) + Zn(s) \rightarrow Zn^{2+}(aq) + Cu$
(c) (i) $Ag^+(aq) + e \rightarrow Ag(s)$ (ii) $Pb(s) \rightarrow Pb^{2+}(aq) + 2e$
(iii) $2Ag^+(aq) + Pb(s) \rightarrow 2Ag(s) + Pb^{2+}(aq)$ (d) (i) $Cu^{2+}(aq) + 2e \rightarrow Cu(s)$
(ii) $H_2(g) \rightarrow 2H^+(aq) + 2e$ (iii) $Cu^{2+}(aq) + H_2(g) \rightarrow Cu(s) + 2H^+(aq)$
(e) (i) $Ag^+(aq) + e \rightarrow Ag(s)$ (ii) $Cu(s) \rightarrow Cu^{2+}(aq) + 2e$
(iii) $Cu(s) + 2Ag^+(aq) \rightarrow Cu^{2+}(aq) + 2Ag$ (f) (i) $Ag^+(aq) + e \rightarrow Ag(s)$
(ii) $Zn(s) \rightarrow Zn^{2+}(aq) + 2e$ (iii) $2Ag^+(aq) + Zn(s) \rightarrow 2Ag(s) + Zn^{2+}(aq)$

19. (a) +0.79 V   (b) −2.37 V   (c) −0.14 V   (d) +1.50 V   (e) −1.66 V
(f) −0.40 V

20. (a) $Zn(s)|Zn^{2+}(aq) \vdots Cu^{2+}(aq)|Cu(s)$ (b) $Mg(s)|Mg^{2+}(aq) \vdots Pb^{2+}(aq)|Pb(s)$
(c) $Al(s)|Al^{3+}(aq) \vdots H^+(aq)|H_2(g), Pt$ (d) $Cu(s)|Cu^{2+}(aq) \vdots Ag^+(aq)|Ag(s)$

10. DEDUCTION OF EQUATIONS (p. 39.)

1. (a) $2Al(s) + 3Cl_2(g) \rightarrow 2AlCl_3(g/s)$ (b) $Zn(s) + I_2(g) \rightarrow ZnI_2(s)$
(c) $2Ag(s) + S(s) \rightarrow Ag_2S(s)$

2. (a) $Zn(s) + 2Ag^+(aq) \rightarrow Zn^{2+}(aq) + 2Ag(s)$
(b) $Mg(s) + Cu^{2+}(aq) \rightarrow Mg^{2+}(aq) + Cu(s)$
(c) $Fe(s) + Pb^{2+}(aq) \rightarrow Fe^{2+}(aq) + Pb(s)$
(d) $2Al(s) + 3Cu^{2+}(aq) \rightarrow 2Al^{3+}(aq) + 3Cu(s)$
(e) $Fe(s) + Cu^{2+}(aq) \rightarrow Fe^{2+}(aq) + Cu(s)$

3. (a) $Mg(s) + 2H^+(aq) \rightarrow Mg^{2+}(aq) + H_2(g)$
(b) $Fe(s) + 2H^+(aq) \rightarrow Fe^{2+}(aq) + H_2(g)$
(c) $2Al(s) + 6H^+(aq) \rightarrow 2Al^{3+}(aq) + 3H_2(g)$
(d) $Zn(s) + 2H^+(aq) \rightarrow Zn^{2+}(aq) + H_2(g)$

4. $CrCl_3$; $CrCl_3(aq) + 3AgNO_3(aq) \rightarrow 3AgCl(s) + Cr(NO_3)_3(aq)$ or
$Cl^-(aq) + Ag^+(aq) \rightarrow AgCl(s)$

5. $TiCl_3$; $TiCl_3(aq) + 3AgNO_3(aq) \rightarrow 3AgCl(s) + Ti(NO_3)_3(aq)$ or
$Cl^-(aq) + Ag^+(aq) \rightarrow AgCl(s)$

6. $CsCl$; $CsCl(aq) + AgNO_3(aq) \rightarrow AgCl(s) + CsNO_3(aq)$ or
$Cl^-(aq) + Ag^+(aq) \rightarrow AgCl(s)$

7. Cr_2O_3; $(NH_4)_2Cr_2O_7(s) \rightarrow Cr_2O_3(s) + N_2(g) + 4H_2O(g/l)$

8. Ag_2O; $2AgNO_3(aq) + 2NaOH(aq) \rightarrow Ag_2O(s) + H_2O(l) + 2NaNO_3(aq)$; 75 cm³

9. $CaCO_3(s) + 2HCl(aq) \rightarrow CaCl_2(aq) + CO_2(g) + H_2O(l)$

10. $Pb(NO_3)_2(aq) + 2NaHCO_3(aq) \rightarrow PbCO_3(s) + CO_2(g) + H_2O(l)$

11. $2CuCO_3, Cu(OH)_2$; $2CuCO_3, Cu(OH)_2(s) \rightarrow 3CuO(s) + 2CO_2(g) + H_2O(g/l)$

12. $Na_2CO_3(aq) + 2HCl(aq) \rightarrow 2NaCl(aq) + CO_2(g) + H_2O(l)$

13. PbO_2; $PbCl_4(l) + 2H_2O(l) \rightarrow PbO_2(s) + 4HCl(aq)$

14. $2AgNO_3(aq) + Na_2S(aq) \rightarrow Ag_2S(s) + 2NaNO_3(aq)$

15. $BaCl_2(aq) + Na_2CO_3(aq) \rightarrow BaCO_3(s) + 2NaCl(aq)$

16. $Pb(NO_3)_2(aq) + K_2CrO_4(aq) \rightarrow PbCrO_4(s) + 2KNO_3(aq)$
17. (i) HgI_2 (ii) HgI_4^{2-}
$Hg^{2+}(aq) + 2I^-(aq) \rightarrow HgI_2(s)$ $HgI_2(s) + 2I^-(aq) \rightarrow HgI_4^{2-}(aq)$

11. DIFFUSION AND EFFUSION OF GASES (p. 43.)

1. 36   **2.** 8:1 (H_2:HI)   **3.** 4:1 (H_2:O_2)   **4.** (a) 36   (b) 72   (c) 36
5. 74.4 s   **6.** (a) CH_2 (b) 28.1   (c) C_2H_4 **7.** 5.7 min   **8.** 18 min
9. 39.8; Ar   **10.** (a) 144   (b) SF_6

12. RATES OF REACTION (p. 45.)

1. (a) 250 cm³   (b) 16 min   (c) 2 g   (d) 0.01 mole   (e) 0.02 mole
(f) (i) 0.4 mol dm⁻³ (ii) 13.6 g dm⁻³   (g) 0.75
2. Rate of reaction is directly proportional to the length of ribbon and thus to its surface
area. 10.8 cm³ min⁻¹
3. (a) 1 min   (b) 0.56 g; 0.285 dm³   **4.** (c) 9.75 cm³ min⁻¹
5. (a) 0–2 min   (b) 4.0 min   (c) 1.75 min   (d) 0.36   (e) all take 4.0 min
(f) 4.0 min
6. (a) 1.3 min   (b) 8.0 min   (c) 6.7 min   (e) 0.86 g   (f) 0.0195 moles
(g) 0.0195 moles   (h) 0.04 moles
(i) $CaCO_3(s) + 2HCl(aq) \rightarrow CaCl_2(aq) + CO_2(g) + H_2O(l)$
7. (a) 0.039 moles   **8.** (a) 9.9 cm³ min⁻¹   (b) 32.0 cm³ min⁻¹
9. (b) 13.35×10^{-3} g min⁻¹ 10.68×10^{-3} g min⁻¹ 8.01×10^{-3} g min⁻¹
5.34×10^{-3} g min⁻¹ 2.67×10^{-3} g min⁻¹   **10.** (i) B   (ii) A   (iii) C

13. SOLUBILITY AND SOLUBILITY CURVES (p. 52.)

1. 74.7 g/100 g water   **2.** (a) 10 g/100 g water   (b) 55 g
3. (a) 23.3 g   (b) 17.7 g   (c) 52.3 g
4. Residue at 70°C: 12 g (all NaCl)   Residue at 15°C: 26 g (1 g NaCl, 25 g KNO_3)
Residue after evapn.: 62 g (37 g NaCl, 25 g KNO_3)
5. (a) 35.5 g   (b) 43 g
6. (a) (i) 95 g (ii) 113 g   (b) 29 g   (c) 57°C   (d) (i) 13 g (ii) 39°C
7. (a) (i) 91.5 g (ii) 119 g   (b) 27.5 g   (c) 35°C   (d) 5.5 g   (e) (i) unsat.
(ii) unsat. (iii) sat. (iv) unsat. (v) sat.
8. (a) 55°C   (b) 14 g/100 g water   (c) 18 g   (d) 14.5 g   (e) 29 g D, 1.6 g E
9. (a) 24°C   (b) all dissolves   (c) 66 g B crystallises   (d) 18.75% A, 81.25% B
10. (a) decreases   (b) (i) 0.285 g/100 g (ii) 0.145 g/100 g   (c) (i) 0.22 g
(ii) 112 cm³   (d) solubility is directly proportional to pressure.